Geosystems. Design Rules and Applications

Geosystems. Design Rules and Applications

A. Bezuijen & E.W. Vastenburg
Deltares, Delft, The Netherlands

CRC Press
Taylor & Francis Group
Boca Raton London New York Leiden

CRC Press is an imprint of the
Taylor & Francis Group, an **informa** business

A BALKEMA BOOK

CRC Press/Balkema is an imprint of the Taylor & Francis Group, an informa business

© 2013 Taylor & Francis Group, London, UK

Typeset by V Publishing Solutions Pvt Ltd., Chennai, India

Published by: CRC Press/Balkema
P.O. Box 447, 2300 AK Leiden, The Netherlands
e-mail: Pub.NL@taylorandfrancis.com
www.crcpress.com – www.taylorandfrancis.com

Library of Congress Cataloging-in-Publication Data

Bezuijen, A.

Geosystems : design rules and applications / A. Bezuijen & E.W. Vastenburg. -- 1st ed.
    p. cm.
  Includes bibliographical references and index.
  ISBN 978-0-415-62148-9 (hardback : alk. paper) -- 1. Geotextiles.
  2. Hydraulic engineering--Equipment and supplies. I. Vastenburg, E.W. II. Title.

  TA455.G44B49 2013
  627--dc23

                                                        2012032876

ISBN: 978-0-415-62148-9 (Hbk)
ISBN: 978-0-203-07286-8 (eBook)

Contents

Preface

Geotextile-encapsulated sand elements are three-dimensional systems manufactured from textile materials that are filled with sand. They form a sub-group of a wider system of geosynthetic solutions for erosion control that are known as Geosystems. These elements are used in hydraulic engineering structures such as dams, dykes and breakwaters as an alternative for quarry stone e.g. as core material. They may also be used for bottom or bank protection or to fill up a scour hole.

The use of these elements has the advantage that local material can be utilized and that no stone needs to be extracted and transported from a quarry to the site. Compared to traditional construction methods (with quarry stone) the application of geotextile sand filled elements may add considerable operational advantages to the execution of hydraulic works and may offer attractive financial opportunities and environmental aspects.

New and promising developments in hydraulic engineering, such as geotextile-encapsulated sand elements, are sometimes hampered by the lack of design rules. That's probably the reason why planners, designers and contractors are therefore rather hesitant to apply these materials.

Most of the design rules are empirical in nature and this means that existing experience is therefore required to design new structures. For this reason it is a challenge to design a structure when only limited experience is available with that type of material.

This was the idea of the CUR-committee F42 (CUR Building & Infrastructure) where existing experiences and research results were gathered and compiled in the CUR-publication 217 "Ontwerpen met geotextiele zandelementen".

The original CUR-217 publication is in Dutch and this manual makes the contents of this publication available for an international audience. Most of this manual is a translation of CUR-217. However, additions have been made to give this manual more international relevance. In addition, some improvements and corrections have been made. When the CUR publication 217 was published the CUR-committee F42 had the following members:

- Dr. J.G. de Gijt (Chairman, Rotterdam Public Works*/Delft University of Technology)
- G.J. Akkerman (Royal Haskoning)
- Dr. A. Bezuijen (Editor, Deltares*/Ghent University)
- E. Berendsen (Ministry of Public Works*)

- J. de Boer (GeopexProducts (Europe) B.V.*)
- C.A.J.M. Brok (Huesker Synthetic GmbH*)
- D. Heijboer (Royal Haskoning)
- M. Klein Breteler (Deltares)
- S. Overal (Ministry of Public Works)
- H.J.W. Ruiter (Van Oord Netherlands n.v.*)
- W.H. Smits (Amsterdam municipality*)
- W.H. Tutuarima (–)
- E.W. Vastenburg (Editor, Deltares/NGO*)
- R. Veldhoen (Van den Herik B.V. Sliedrecht*)
- D.P. de Wilde (Ministry of Public Works)
- E.L.F. Zengerink (Ten Cate Nicolon B.V.*)
- K.W. Pilarczyk (Ministry of Public Works)
- A. Jonker (CUR, coordinator)

* Financiers CUR publication 217.

For this version, the translation was made by translation company CPLS. This translation was edited by E.W. Vastenburg and Dr. A. Bezuijen. Final editing and quality scan was done by C. Lawson. In addition, technical changes are discussed with the members of the F42 committee.

Research for this publication and translation in English was funded by Delft Cluster and Deltares. The Dutch Chapter of IGS, NGO, participated in the research and the translation work for this publication.

Notations

A	Filled cross-sectional area of the geotextile-encapsulated sand element (perpendicular to the longest axis)	m²
A_g	Flow surface	m²
A_s	Cross-sectional area in the horizontal plane	m²
B	Exponent dependent on the interaction between waves and structure	–
b	Width of structural element (geotextile element)	m
b_m	Width of the geotextile mattress	m
b_g	Width of the geotextile	m
b_f	Thickness of the filter layer	m
b_k	Thickness of the clay layer	m
b_0	Maximum width of the opening of the split barge	m
B_{tot}	Total width of the layer of the geotextile containers under consideration	m
c	– Concentration of the sand in the mixture	m³/m³
	– Constant	–
c_v	Consolidation coefficient of the grain skeleton	m²/s
C_d	Drag force coefficient	–
C_u	Uniformity coefficient	–
d	Drainage distance	m
D	Diameter of the geotextile tube at 100% filling	m
D_d	Maximum diameter of the tubes or sausages (in the case of geotextile mattresses)	m
D_{eq}	Equivalent thickness of the structural element	m
D_{mf}	Average grain diameter	m
D_n	Nominal thickness of a geotextile bag	m
D_k	Effective thickness of the geotextile-encapsulated sand element	m
D_x	Sieve size through which x% fraction of the sand material passes	m

D_t	Thickness of the shear-susceptible layer of the geotextile containers	m
E	Modulus of elasticity of the geotextile	kN/m² or N/mm²
E_{geo}	Maximum energy to be dissipated by the geotextile per unit of length	J/m
E_{fall}	Fall energy per unit of length	J/m
E_{fill}	Maximum energy to be dissipated by the fill material per unit of length	J/m
f	Degree of filling with respect to the cross sectional area (percentage of the area of the theoretical circle)	–
f_m	Mobilised surface friction coefficient between adjacent geotextile containers	–
F_{rac}	Ratio of the fall energy to the maximum of E_{fill} and E_{geo}	–
F	Stability factor	–
F_o	Outward force created by the hydraulic head perpendicular to the slope per unit length	N/m
g	Gravitational acceleration	m/s²
G	Submerged weight of a layer of containers per unit length	kg/m
h	– Height of the geotextile-encapsulated sand element	m
	– Water depth	m
H	Height	m
H_{pile}	Height of the pile	m
h	Filled height of the geotextile tube	m
H_s	Significant wave height	m
i	Hydraulic gradient in the fill material	–
J	Tensile stiffness of the geotextile	kN/m or N/m
k_s	Permeability of the fill material	m/s
k_g	Permeability perpendicular to the geotextile	m/s
k_r	Nikuradse roughness	m
K_T	Turbulence factor	–
K_h	Factor related to the depth	–
K_s	Factor related to the slope	–
L	Length of the geotextile-encapsulated sand element	m
l	Length of the geotextile mattress	m
L_0	Wave length in deep water	m
L_{bw}	Length of the geotextile mattress above water	m
L_{ow}	Length of the geotextile mattress under water	m
M	Mass of the geotextile-encapsulated sand element	kg
n	Porosity of the fill material of the structural elements	–
O_{90}	Pore size of the geotextile that corresponds to the average diameter of the sand fraction of which 90% remains on the geotextile (in the wet sieving method)	m
p	Pressure in the fill material	kN/m²
q	Specific discharge	m/s

Q	Sand production	m³/hr
R_c	Distance between the crest and still water line	m
r	Radius of curvature at a random point on the geotextile skin	m
s_{op}	Wave steepness	–
S	Circumference of the cross-section of the geotextile-encapsulated sand element (perpendicular to the axis)	m
t_g	Thickness of the geotextile	m
τ'	Characteristic time	–
T	Tensile load in the geotextile	kN/m or N/m
T_d	Characteristic drainage period	s
T_n	Characteristic compression period	s
T_m	Maximum allowable tensile load in the geotextile	kN/m or N/m
T_{seam}	Tensile load of the seam	kN/m
T_p	Peak period	s
u_{cr}	Critical horizontal flow velocity along the surface of the structure	m/s
v	Fall velocity of the geotextile-encapsulated sand element	m/s
v_{sed}	The sedimentation rate (the rate at which the sand bed rises)	m/s
v_{cr}	Pump speed	m/s
v_∞	Terminal fall velocity	m/s
V	Volume of the geotextile-encapsulated sand element	m³
z	Depth of the subsoil	m
w_0	Fall velocity of a single grain	m/s
α	Slope angle of the structure	deg
α_c	One-dimensional compressibility of the grain skeleton upon discharge	m²/kN
β	Angle of shear of the geotextile-encapsulated sand elements	deg
X	Shape factor according to Van Rhee [29]	–
γ	Overall safety factor	–
δ	Friction angle between geotextile element and the subsoil	deg
Δ_t	Relative density of the structural elements/relative density of the granular filter	–
Δ	Relative density of the fill material	–
Δ_s	Relative density of the subsoil	–
Δ_n	Porosity reduction under constant wave load	–
$\Delta\phi$	Difference in piezometric head over the geotextile	m
ε_m	Maximum strain of the geotextile	%
ε_{max}	Strain at maximum allowable tensile strength of the geotextile	%
ϕ	Angle of internal friction of the type of encapsulated sand element	deg
ϕ_s	Angle of internal friction of the fill material	deg
φ	Numerical constant	–
Φ	Stability parameter	–
ρ	Density of the geotextile-encapsulated sand element	kg/m³
ρ_s	Density of the fill material (mass of the grains)	kg/m³
ρ_{mat}	Density of the saturated geotextile-encapsulated sand element	kg/m³
ρ_w	Density of water	kg/m³

ρ'	Buoyant density of the geotextile container	kg/m^3
σ_i'	Average effective stress of the sand in the container	kN/m^2
s_p	Standard deviation of the placement accuracy	m
Φ	Expected maximum hydraulic head difference over the outermost layer of the geotextile containers	m
ξ	Breaker parameter	–
Ψ	– Shields parameter	
	– Permittivity	s^{-1}
	– Dilatancy angle	deg
Ψ_0	Generation of excess pore pressure for undrained load	$kN/m^2 s$
Ψ_u	Valuation factor dependent on type and quality of material and structure	–
ν	Kinematic viscosity	m^2/s

List of figures

Pictures/figures marked with the annotation 1, 2, 3 or 4 (in superscript) have been provided by:

1   TenCate Geosynthetics Group
2   Geopex-Profix
3   Deltares
4   Arne Bezuijen

Chapter 1

Introduction

1.1 WHY THERE IS A NEED FOR DESIGN RULES FOR GEOTEXTILE-ENCAPSULATED SAND ELEMENTS

Geotextile-encapsulated sand elements are three-dimensional systems manufactured from geotextiles (woven and/or nonwoven materials) filled with sand. These elements can be regarded as innovative, (sometimes) economic and environment-friendly systems for hydraulic structures along inland and coastal waters, where they form an alternative to the use of traditional hydraulic materials. There are four distinct types of these elements, namely geotextile bags, geotextile mattresses, geotextile tubes and geotextile containers. Table 1.1 provides a summary of the main areas of application for each type.

Experience with these structures has been obtained in many regions, for example, Australasia, South East Asia, South Asia, Europe, Africa, North and South America. However, there is the impression that broader application is possible if knowledge of the performance of geotextile-encapsulated sand elements is more readily available. An initial step has already been taken in The Netherlands with [14] to make current knowledge of geosystems more accessible to a broader public, so that they can be incorporated in the planning stage as a structure variant. Furthermore, an extensive description of the available knowledge is provided by Pilarczyk [22]. Based on the work presented by Pilarczyk and other sources (see the bibliography) this manual will provide design rules for the most common geotextile-encapsulated sand elements. Where possible, the design rules have been validated and, where appropriate, the limitations stated. The target group for this manual comprises project managers, designers, administrators and supervisory bodies involved in the planning and execution of hydraulic and coastal structures.

1.2 DOCUMENT STRUCTURE

This manual contains a summary of design rules and calculation examples for geotextile-encapsulated sand elements based on the available theories and describes the standard construction methods for these types of structures.

Chapter 2 describes the general design approach applied to these geosystems. In addition to the general design approach to be followed for all hydraulic structures, the safety aspects and the required properties of geotextiles are considered. Chapters 3 to 6 examine in more detail the various applications and element types; geotextile bags (3), geotextile mattresses (4), geotextile tubes (5) and geotextile containers (6). Details covered include filling methods, applied loads, required strengths, stability, durability and positioning accuracy. Examples are given along with pointers for practical application.

Table 1.1 Various geotextile-encapsulated sand elements and their applications.

Example of elements	*Possible applications*

- **Geotextile bags**

  Coastal applications: beach groyne; breakwater; dune toe protection; channel repair; soil and bank protection; dyke closure.

  River and stream applications: submerged breakwater; groyne and sediment management; soil and bank protection; scour hole repair.

  Other: nature development areas.

- **Geotextile mattresses**

  River and stream applications: soil and bank protection.

  Other: nature development areas.

- **Geotextile tubes**

  Coastal applications: groyne; breakwater; dune foot protection; submerged revetments; channel repair; land reclamation; artificial reef; sill structures.

  River applications: submerged breakwater; groyne, sediment management; river training; bank protection.

  Other: nature development areas; dewatering of dredged material; temporary structures.

- **Geotextile containers**

  Coastal applications: breakwaters; sill structures; channel repair; land reclamation; artificial reef; dyke closure; toe stability.

  River applications: submerged breakwater; groyne and sediment management; scour hole repair.

  Other: storage of dredged material; temporary works.

Chapter 2

General design aspects

This chapter considers the design methodology of structures using geotextile-encapsulated sand elements. First a description of the principles of geosystems is given, then 2.2 describes the design process for these geosystems. The safety philosophy is then examined (2.3), with the importance of identifying potential failure mechanisms being crucial to the design process emphasized in 2.3.3. Finally, 2.4 details the different material aspects of geotextile containers.

2.1 BASIC PRINCIPLES OF GEOTEXTILE-ENCAPSULATED SAND ELEMENTS

Sand is commonly used in hydraulic structures. The main reasons being that sand is one of the cheapest building materials available, can be obtained in large quantities, is easy to process and is reusable. Sand is also mechanically stable, volumetrically stable, and has predictable engineering properties. The drawback with sand, however, is that it lacks cohesion and erodes easily under the influence of current and waves; and when dumped under water, it tends to migrate even at gentle slopes. To build a sand structure in water, it is normally necessary to contain it by creating a fore-bank (for a quay structure) or a containment dyke. These containment structures are normally composed of gravel, mine stone, slag or armour stone. These coarse materials are also able to fulfil an additional function – protection of the sand from erosion by currents and waves. Filter layers are often required to prevent the sand from eroding through the pores of the larger-sized, granular material.

Working with stone-like materials can be expensive. Geotextile-encapsulated sand elements are an attractive alternative since local materials can be used and (less) armour stone needs to be extracted and transported, which presents (depending on local prices) a possibility to reduce cost. In addition, a relatively steep, erosion resistant slope can be constructed in water using these elements that are comparable with armour stone. Both the production of these geosystems and the construction of the structure in water are relatively straightforward. If required for temporary purposes only, the structure may also be easily removed.

One disadvantage of geotextile-encapsulated sand elements is that they are susceptible to damage during installation. Damage may occur due to poor installation

Table 2.1 Application areas of geotextile-encapsulated sand elements.

Application	Geotextile bag	Geotextile mattress	Geotextile tube	Geotextile container
Beach groyne	X		X	
Breakwater	X		X	X
Hanging beach	X		X	X
Dune toe protection	X		X	
Channel repair	X	X	X	X
Land reclamation			X	X
Underwater reef	X			X
Bed protection	X	X		
Bank protection	X	X	X	
Temporary dams	X		X	X
Sediment management	X	X	X	X

practice, the uncontrolled placement of armour stone against the geotextile skin, and from vandalism of the exposed geotextile skin. Due to these susceptibilities, and because geotextiles lose strength over time when exposed to ultra-violet light, permanent structures utilizing these geosystems are almost always covered with one or more layers of granular material. Depending on the application, and the type of geotextile used, a cover layer may be omitted for temporary structures.

For large elements (e.g. geotextile tubes) no damage should occur to the geotextile skin as this will allow the sand-fill to be lost. For small elements (e.g. geotextile bags) limited damage can be sustained provided there are sufficient surrounding elements in the structure that are not damaged.

These geosystems may also be used in combination with materials other than sand-fill (for example sludge). Applications exist where fine soils are used in combination with geotextile encapsulation to drain and prevent their erosion. This manual, however, focuses exclusively on sand-filled encapsulation.

The appendices of [14] describe a number of geotextile containment projects already carried out. Table 2.1 contains a brief summary of the application areas involved.

2.2 DESIGN APPROACH

The construction of a hydraulic engineering project comprises several phases, from initiative to aftercare. Each phase generates an end-product that allows the next phase to begin. In chapter 2 of [11] and chapter 2 of [22] this process is described in detail. This manual is specifically focused on the planning phase (i.e. the design process), and is intended to provide guidance for the rational design of geotextile-encapsulated sand element structures.

Irrespective of the contract form chosen, this manual can be used as an aid to generate both the Programme of Requirements (specification of functional requirements for innovative contracts) and the design of a structure using these geosystems. Since the design (type of geotextile, and type and size of elements and their geometrical layout) is governed to a significant extent by the choice of materials to be used and the corresponding construction method, a successful design requires close interaction between the designer, supplier of geotextiles and the installation contractor.

The design of geotextile-encapsulated sand elements is characterised by an iterative process (as is usual in design processes), where various solutions are sought that comply with the functional and performance requirements. This manual assumes that the designer has already gone through the initial phase of the design process; the initiation and the exploration phase, and has opted for a design using these geosystems. The design process begins with the chosen type of geotextile-encapsulated sand elements (which has normally been chosen on the basis of design data, e.g. water depth, dimensions of structure, hydraulic loading, etc). Figure 2.1 illustrates this design process.

Based on the functional requirements and the local conditions (specifically the hydraulic and geotechnical conditions), the key dimensions of the overall structure are established, such as the characteristic cross-section, longitudinal profile, crest height and the slope. Further, the construction procedure is taken into account and this should include an inventory of the available fill material.

Next, the specific dimensions and component configuration of the structure can be determined using, for instance, applications and practical examples given in the literature [46 and 47]. The size (weight and dimensions) of the geotextile-encapsulated sand elements and their number, can then be determined.

When the overall size and dimensions of the structure are known the potential failure mechanisms can be established and the interrelationship between these failure mechanisms may be drawn up in the form of a fault tree (see 2.3.3).

The next step is to assess the hydraulic stability of the elements – the impact of waves and/or currents. In many cases the geotechnical stability of the elements must also be assessed, e.g. when they are used as bank protection. This stability assessment must account for both the construction and the operational phases. The construction phase is often dominant. If the elements do not fulfil the stability requirements during construction, then a different size (weight/dimension) must be chosen. This does have an impact on the number of elements to be used and possibly on the method of construction.

If the elements fulfil the various stability requirements, the minimum tensile strength and strain capacity needed for these elements can then be determined. This minimum tensile strength also applies to the required strength capacity of any seams (often the weakest part in these elements). From this, the required tensile strength and strain capacity for the geotextile can be established, followed by the choice of material for the geotextile. Finally, the chosen geotextile is checked for durability and for "blocking" and "clogging" resistance (see Appendix A).

In principle, this general approach applies to all the geotextile-encapsulated sand elements considered in this manual, where the calculations for determining the strength of the geotextile are related to specific, proven, methods of construction. Chapters 3 to 6 cover the general design procedure for each of the various types of these geosystems.

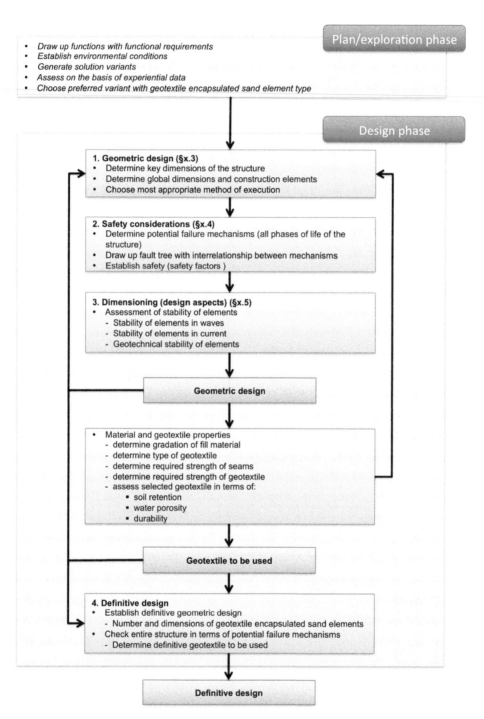

- *Draw up functions with functional requirements*
- *Establish environmental conditions*
- *Generate solution variants*
- *Assess on the basis of experiential data*
- *Choose preferred variant with geotextile encapsulated sand element type*

Design phase

1. Geometric design (§x.3)
- Determine key dimensions of the structure
- Determine global dimensions and construction elements
- Choose most appropriate method of execution

2. Safety considerations (§x.4)
- Determine potential failure mechanisms (all phases of life of the structure)
- Draw up fault tree with interrelationship between mechanisms
- Establish safety (safety factors)

3. Dimensioning (design aspects) (§x.5)
- Assessment of stability of elements
  - Stability of elements in waves
  - Stability of elements in current
  - Geotechnical stability of elements

Geometric design

- Material and geotextile properties
  - determine gradation of fill material
  - determine type of geotextile
  - determine required strength of seams
  - determine required strength of geotextile
  - assess selected geotextile in terms of:
    - soil retention
    - water porosity
    - durability

Geotextile to be used

4. Definitive design
- Establish definitive geometric design
  - Number and dimensions of geotextile encapsulated sand elements
- Check entire structure in terms of potential failure mechanisms
  - Determine definitive geotextile to be used

Definitive design

Figure 2.1 Iterative design process for geotextile-encapsulated sand elements.

* The initiative/exploration phase falls outside the scope of this design manual.

2.3 SAFETY CONSIDERATIONS

2.3.1 Introduction

From a design perspective, the structure must meet its functional requirements over the design lifetime. This also applies to the probability of failure, which must be lower than the pre-established target, which is sometimes defined per failure mode. This means that the probability of a load on the structure exceeding its resistance remains below a target value, which itself is established considering the potential consequence of structural failure. The use of probabilistic reliability concepts is becoming more and more popular in the civil engineering branch lately. There are four distinct levels:

- Level III: Fully probabilistic method;
- Level II: Semi-probabilistic method;
- Level I: Quasi-probabilistic method (partial safety factors);
- Level 0: Deterministic method (overall safety factor).

These methods are described, and used, in detail in [10].

2.3.2 Deterministic method

For designing with geotextile-encapsulated sand elements the knowledge and experience has not yet developed to the extent that probabilistic methods can be used reliably. Therefore, this manual focuses on deterministic methods only. For loads and strengths, fixed design values are incorporated for each parameter separated by an overall safety factor.

$$R \geq \gamma \cdot s \tag{2.1}$$

where:

R = resistance (strength);
s = loads;
γ = overall safety factor.

The overall safety factor used must have a value greater than or equal to 1.0 (in practice an overall safety factor between 1.1 and 1.5 is common). In many cases there are no formal procedures available to determine the appropriate overall safety factor. In practice, the established safety is that which is still available in respect of the (characteristic) design value. It is thus visible to both the designer and the user what safety is present in the structure in relation to the design values. This safety approach is adopted in this manual.

Strictly speaking, overall safety factors are used to compensate for uncertainties and coincidences. The method is also applied when taking account of the reduction of the strength of the geotextile material, caused by UV radiation, thermal oxidation, seams, creep or damage during construction. There is a close relationship with the question: what is the desired (economic) lifetime, and when may a loss of structural function occur? The overall safety factor thus comprises a collection of partial safety factors that all reduce the design value of the strength of the geotextile. In 2.4 these partial factors are examined more closely. Fixed partial factor values are difficult to provide because they are dependent on the required economic lifetime. Currently, the following partial factors for the material are being used: weakening by seams 2, damage during construction 1.25, loss of strength through creep 1.4 to 4 (dependent on material type). These default values must be used with caution, with a minimum overall safety factor of 3.5 to 4.0 being applied.

The stability relationships used in this manual for the loads, e.g. waves, currents, etc., are empirical in nature based mainly on model research. For these relationships, safety is implicitly included since a stability relation, based on the performed research, will be drawn underneath all points where instability was measured. Therefore, no additional safety factors to these stability relationships have been applied. Thus, the application of safety factors here only relates to the strength properties of the geotextile to be used. It should be noted that this is only valid for the conditions tested in the performed research.

The designation of the appropriate safety factors for the different geotextile-encapsulated sand elements are detailed in the respective sections entitled "Failure mechanisms and safety considerations" of chapters 3, 4, 5 and 6.

2.3.3   Potential failure mechanisms

The designer must have an understanding of the potential failure mechanisms that could occur in hydraulic structures. Experience has shown that whenever a structure fails, it tends not to be as a result of underestimating the loading but is more often due to a failure mechanism that was not considered during design. It is, however, outside the scope of this manual to detail all possible potential failure mechanisms for different hydraulic structures, e.g. wave overtopping, piping and geotechnical instability during the construction and in-situ phases. We refer to the various existing publications such as [11].

The potential failure mechanisms considered here all relate to the grouping (with the exception of geotextile mattresses) of geotextile bags, geotextile mattresses, geotextile tubes and geotextile containers, as used in hydraulic structures. The individual chapters in this manual examine these potential failure mechanisms for each type of element, related to both the construction and operational phases, such as:

- instability due to wave and current loadings;
- instability due to tilting, rolling, slipping or overturning;
- washing out of fill material through the pores of the geotextile;
- puncture of the geotextile as a result of the impact of sharp-edged armour stone;
- failure of seams and/or geotextile through tensile rupture;
- liquefaction of sand in the structure leading to adverse deformation.

The cause and effect of each potential failure mechanism, as well as the interrelationship with other potential failure mechanisms can be illustrated by a fault tree that defines an undesired event, e.g. the failure of a bank protection made of geotextile bags. The branches of the fault tree show the various causes that could lead to this event. The probability of the occurrence of these individual potential failure mechanisms and the relationship between them generates the probability of this undesirable

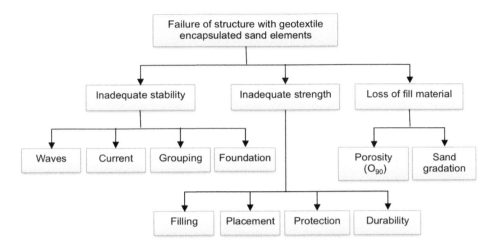

Figure 2.2 Fault tree for a structure with geotextile-encapsulated sand elements.

event occurring. Figure 2.2 contains an example of an appropriate fault tree for geotextile-encapsulated sand elements in a hydraulic structure.

2.4 MATERIAL ASPECTS OF GEOTEXTILES

2.4.1 Raw materials

When making a design using geotextile-encapsulated sand elements, the properties and behaviour of the geotextile play a crucial role. This section provides an overview of the key properties of the geotextiles used in Dutch geotechnical, road and hydraulic engineering works. Much of this information has been taken from [12, in Dutch] and [22], in which much knowledge related to the application of geotextiles in hydraulic engineering is clustered.

"Geotextile" is the collective name for both woven and nonwoven materials. Woven geotextiles are formed from yarns or tapes into an ordered 2-dimensional structure. Nonwoven geotextiles comprise continuous or staple fibres that are randomly oriented and form a stable structure due to mechanical (needle-punching), thermal or chemical bonding. For geotextiles the following raw materials are used:

- polyester (PET);
- polypropylene (PP);
- polyethylene (PE).

Table 2.2 lists some characteristic properties of the three raw materials used for geotextiles.

In civil engineering PET, PP and, to a lesser extent, PE materials are used for geotextiles. In environmental engineering it is mainly PE, and PET to a lesser extent, due to the often prevailing chemical load.

Table 2.2 Some properties per raw material for geotextiles.

Material	Unit weight [kg/m³]	Tensile strength [N/mm²]	Elastic modulus [N/mm²]	Maximum strain [%]
Polyester (PET)	1380	800–1200	12000–18000	8–20
Polypropylene (PP)	920	400–800	2000–8000	6–25
Polyethylene (PE)	900–930	350–600	600–6000	10–30

PP and PET are the most commonly used materials for application in geotextile-encapsulated sand elements, fulfilling separation, filtration and containment functions. If the geotextile remains under tension in the structure, the use of polyester could prove advantageous because of its better long-term load carrying capabilities.

When using PP and PE under water, account must be taken of the specific gravity of these materials (between 0.9 and 0.95), which is lower than that of water, so the materials will float.

The following sections deal with the sand tightness, the permeability and the strength, and following this is an appraisal of the quality of the seams, damage during installation and the durability of geotextiles. Finally, the main aspects of the CE (Conformité Européene) Marking are stated.

2.4.2 Sand tightness

When using geotextiles for filtration purposes, to prevent the erosion of fine particles through the geotextile in hydraulic structures a distinction is made between geometrically open and geometrically closed geotextiles. The pores in geometrically closed geotextiles are so small that the sand is physically blocked, so it will prevent erosion for all hydraulic gradients. Geometrically open geotextiles act as a filter over a limited gradient range. At higher gradients the sand could be washed out.

In geotextile-encapsulated sand elements only geometrically closed geotextiles are used because high gradients can occur. The basis for determining the required pore size (O_{90}) of the geotextile is that only a very small amount of sand may be allowed to erode, otherwise the encapsulated sand elements will deform too much. Table 2.3 shows the recommended design criteria for sand retention [22].

During the filling of geotextile-encapsulated sand elements there will also be a dynamic load if the elements are filled hydraulically. Some initial loss of sand is not serious during the filling process, however, there must be no loss of fill material over time.

The above retention criteria are conservative and are based on an uncovered encapsulated sand element with a high hydraulic load. For a steeper sand gradation curve $(C_u < 2)$, and moderate load, the criteria can be relaxed to some extent, with the size of the openings a little larger, for example, $D_{90} < O_{90} < 2 \cdot D_{90}$. Large-scale experiments in the Delta flume of Deltares have revealed that these criteria are not valid when the sand can still move around in the structure behind the geotextile. The criteria presented in Table 2.3 assume there is a certain confining stress exerted

Table 2.3 Recommended design retention criteria for geometrically closed geotextiles.

	Sand (D > 60 μm)
Stationary hydraulic load (current)	$O_{90} < 5\,D_{10}\,C_u^{1/2}$ and $O_{90}\,2\,D_{90}$
Dynamic hydraulic load (wave attack)	$O_{90} < 1.5\,D_{10}\,C_u^{1/2}$ and $O_{90} < D_{90}$

O_{90} = pore size of the geotextile that corresponds to the average diameter of the sand fraction of which 90% remains on the geotextile (in the wet sieving method).
D_x = sieve size through which x% fraction of the sand material passes.
C_u = uniformity coefficient of the sand (D_{60}/D_{10}).

on the sand grains. In this case, washing out of fines will be limited because the sand acts as a natural filter. However, if the geotextile can move and there is not always a confining stress present a natural filter cannot be formed and hence a considerable amount of fine particles can pass through the geotextile. For such a situation stricter retention criteria are necessary as presented by Heibaum [35].

2.4.3   Permeability

The permeability of a geotextile is of limited significance for the filling operation of geotextile-encapsulated sand elements unless hydraulic fill is used. In hydraulic structures, geotextile permeability can, however, be important. The permeability of the geotextile must not hinder groundwater flow through the structure, e.g., in slope protection where inadequate permeability can lead to excess pore pressures that result in possible instability of the slope.

A commonly used empirical relationship is that the permeability of the geotextile, as measured by an index test (see 2.4.8), must be a factor of 10 times higher than the permeability of the soil being protected. A more detailed discussion of geotextile permeability and its relevance to soil filtration is given in Appendix A.

Application of the empirical relationship that the permeability of the geotextile has to be 10 times the soil permeability is sometimes awkward given that the permeability of the geotextile is a function of its thickness, and the thickness of the geotextile incorporated in a structure is often unknown. In other instances there may be little hydraulic pressure drop through the geotextile. Consequently, use is then made of the permittivity property of the geotextile, see also Appendix A.

2.4.4   Tensile strength and strain

Normally the highest load is exercised on geotextile-encapsulated sand elements during the construction phase. The tensile strength of the geotextile and the strength of the seams must be sufficient to resist the loads imparted during the filling, transporting and placement of the elements. Depending on the application and construction method, the tensile strength, strain and the base geotextile material used can all be important considerations. Woven geotextiles generally have a relatively high tensile strength and a low maximum strain, while nonwoven geotextiles have a relatively low tensile strength and a high maximum strain. The Young's modulus E or the tensile stiffness J of the geotextile can be derived from its maximum allowable tensile load (T_m) and corresponding strain.

$$E = \frac{T_m}{\varepsilon_m \cdot t_g} \tag{2.2}$$

$$J = E \cdot t_g = \frac{T_m}{\varepsilon_m} \tag{2.3}$$

where:

t_g = thickness of the geotextile [m];
ε_m = maximum strain of the geotextile [−].

T_m is defined as the maximum allowable tensile load in the geotextile, equal to the maximum tensile strength of the geotextile.

It should be realised that, in reality, there is normally a non-linear relation between the tensile load and the strain and hence the stiffness modulus derived in this way is only an approximation of the real stiffness behaviour.

Table 2.4 lists indicative tensile strengths and corresponding maximum strains of the most common woven geotextile types. For the calculation of J use is made of the average value of ε_m.

2.4.5   Seams

The overall strength of these elements is in many cases governed by the strength of the seams. The seam strength depends on the type of seam, the thread type and stitch density used, and can vary from 30% to more than 80% of the strength of the geotextile. Table 2.5 shows the most commonly applied seams and the corresponding typical strength efficiencies obtained. This table is derived from Table 8 in [8] and relates to seams made in the factory. For seams made on-site, by a hand sewing machine, lower strengths are normally obtained. For a stronger (heavier) geotextile this percentage

Table 2.4 Tensile strength and corresponding strain for various various geotextile types [22].

Geotextile	T_m [kN/m]	ε_{max} [%]	J [kN/m]	$T_m * \varepsilon_m$ [kN/m]
Polyester (PET)	100–1600	8–15	870–16000	8–210
Polypropylene (PP)	40–300	10–15	320–2400	4–45
Polyethylene (PE)	20–50	20–30	80–200	4–15
Membrane	7–70	50–100	9–90	4–70

T_m = tensile strength of the geotextile [kN/m].
ε_m = strain at maximum allowable tensile strength of the geotextile [%].
J = tensile stiffness modulus of the geotextile [kN/m].

Table 2.5 Type of seams with typical strength efficiencies for geotextiles.

Seam	Description	Strength of seam on seam side
	Prayer seam common in geotextile containers when closure is made on-site	30–50% of strength of geotextile
	Butterfly seam	40–70% of strength of geotextile
	J-seam	30–60% of strength of geotextile
	Double J-seam	50–70% of strength of geotextile
	Overlap Z-seam can only be produced in the factory with special equipment	>80% of strength of geotextile

reduction is even larger. For geotextile containers on-site closures have been developed with strengths of up to 80% of the strength of the geotextile.

In this manual it is assumed (e.g. in the calculation examples) that the seams have been properly stitched. If this is not the case, then the maximum tensile load capability will decrease correspondingly. For quality control purposes, it is necessary for the seam strength to be controlled during the manufacturing process with subsequent traceability.

2.4.6   Damage during installation of gravel layers

For all applications where geotextiles are used, and thus also for applications with geotextile-encapsulated sand elements, the installation process is critical in respect to installation damage. Distinction can be made between major installation damage (tearing, weakening of seams, etc.) which can be attributed to design and construction errors, and minor installation damage. An example of minor installation damage is where a covering layer of armour stone on a geotextile may result in the geotextile being punctured at various locations and this may lead to a loss of fill material. In these cases it may be desirable to apply a cushion layer of smaller gravel material directly on the geotextile prior to applying the armour stone. It is important that the diameter of the stones used directly on top of the geotextile be sufficiently small, resulting in limited point loads on the geotextile-encapsulated sand elements (more information on this can be found in [11]).

2.4.7   Durability

The lifetime of structures constructed of these geosystems is mainly governed by the durability of the geotextile used. The properties of geotextiles may change over time due to:

- UV radiation;
- oxidation;
- hydrolysis;
- chemical and biological exposure;
- mechanical damage;
- creep and/or relaxation.

In Table 2.6 a qualitative summary is provided of the inherent resistance of geotextiles to various kinds of exposure.

As geotextiles age, oxidation can occur through a combination of temperature, UV radiation and environment, and over time this can result in the geotextile becoming brittle. To combat this, anti-oxidants and UV stabilisers are added to the basic material of the geotextile. These govern the long term quality of the geotextile.

When polyester geotextiles are placed in a hydraulic environment a chemical reaction known as hydrolysis may reduce their strength over time. The rate of hydrolysis is very slow, but it may be accelerated at increased pH levels. For geotextiles manufactured from high molecular weight, high modulus polyester polymer the rate of strength reduction due to hydrolysis is negligible in hydraulic conditions (where there is no UV exposure).

Chemicals and bacteria in the vicinity of the geotextile may also be detrimental in certain circumstances. Polyester is, for example, susceptible to a strong alkali environment and polypropylene is sometimes adversely affected by strong oxidising agents.

Another time-associated phenomenon is creep, where the geotextile continues to deform under an applied load. This deformation can be such that the material ultimately fails. In Table 2.7 the modified allowable tensile strength allowing for the effects of creep is shown for various geotextile materials. Polypropylene (PP) and polyethylene (PE) can weaken due to creep at medium tensile stresses after a relatively short time period. Polyester shows good creep resistance and thus may be used effectively for long-term load carrying applications. However, for applications using geotextile-encapsulated sand elements tensile loads only occur during element

Table 2.6 Resistance of geotextile raw materials to various environments. Since LDPE is not used in geotextile-encapsulated sand elements the properties of HDPE are shown here [22, paragraph 3.1.3].

Raw material	Polyester		Polypropylene		Polyethylene	
Exposure duration	Short	Long	Short	Long	Short	Long
Diluted acids	++	+	++	++	++	++
Concentrated acids	0	−	++	+	++	++
Diluted alkali	++	0	++	++	++	++
Concentrated alkali	0	−	++	++	++	++
Salt	++	++	++	++	++	++
Mineral oil	++	++	+	0	+	0
Glycol	++	0	++	++	++	++
Micro-organisms	++	++	++	++	++	++
UV-light	+	0	0	−	0	−
UV-light*	++	+	++	+	++	+
Dry heat up to 100°C	++	++	++	+	+	0
Steam up to 110°C	0	−	0	−	+	0
Liquid absorption	++	++	++	++	++	++
Detergents	++	++	++	++	++	++

Short = from manufacture until installation.
Long = during actual lifetime of the works.
*   = resistance after addition of UV stabilisers.
++   = good resistance; + = reasonable resistance; 0 = poor resistance; − = no resistance.

Table 2.7 Allowable tensile strength as percentage of the original tensile strength due to creep at 20°C.

Material	Percentage of maximum tensile strength due to creep		
	2 years	10 years	50 years
Polyester	75%	70%	60%
Polypropylene	50%	40%	25%
Polyethylene	50%	40%	25%

placement. Once the element has been placed, the applied tensile loads tend to be small and thus the effects of creep are negligible.

When using polyester geotextiles one should be aware that during the filling with sand the filaments of the geotextile may become damaged due to abrasion by the movement of sand. Other geotextile types are less susceptible to abrasion.

In summary, in selecting the geotextile the following aspects require attention:

- duration of the stress during the life of the structure;
- manner of construction;
- duration of the exposure to UV radiation;
- rate of leaching of anti-oxidants and UV stabilisers;

- presence (in the ground or in water) of metals that may act as a catalyst for the ageing process;
- aggressiveness of the environment (presence of organisms in the soil, acid or alkali environment, etc).

For most applications of geotextiles it is the manner of construction and the extent of exposure to UV radiation that is important. As is shown in this manual, geotextile-encapsulated sand elements are subject to tensile loads during installation only, and thus, little account needs to be taken of longer-term mechanical effects.

2.4.8   CE marking

Many testing methods have been developed for geotextiles, several of which have been standardised by European (EN) or International (ISO) standards. In the context of CE

Table 2.8 Sample standard specifications of geotextiles.

Geotextile		GT 750M	GT 1000M	PE	Non-woven (usually PP)
Structure					
Structure type		Textile materials/ fabric	Textile materials/ fabric	Textile materials/ fabric	Nonwoven
Yarn type		Split fibre	Split fibre	Monofilament	–
Colour		Black	Black	Black	–
Mechanical properties					
Nominal tensile strength MD	kN/m	120	200	40	19–28
Strain at nominal tensile strength	%	15	15	25	50–80
Tensile strength at 10% strain	kN/m		–	–	–
Tensile strength at 5% strain	kN/m	30	55	–	–
Tensile strength at 2% strain	kN/m	–	–	–	–
Nominal tensile strength CD	kN/m	110	200	35	–
Strain at nominal tensile strength	%	11	12	25	–
Puncture strength	kN	14	20	4	4
Displacement upon puncture	mm		45	40	–
Cone drop test	mm	6	6	12	–
Hydraulic properties					
Permeability (at 50 mm)	l/m²/s	20	20	500	60
Water column at 10 mm/s	mm	30	30	7	–
Pore size (O_{90})	µm	250	320	525	90
Durability					
UV resistance	klasse	C	C	B	–
Oxidation resistance	klasse	B	B	D	–
Physical properties					
Weight	g/m²	570	840	210	500
Thickness	mm	1,0	1,4	0,8	3,6
Reel width	m	5,20	5,20	5,05	5,0
Reel length	m	200	200	100	50
Reel diameter	m	0,5	0,6	0,30	0,5
Reel weight	kg	380	480	105	135

Marking, manufacturers in Europe are obligated, depending on the application of the geotextile, to report the relevant test properties for each geotextile, and to ensure that the geotextile supplied also conforms to the properties stated on the label. However, these standardised tests in the context of European legislation are index tests only. The properties show results for the tested piece of geotextile only and do not account for interaction with the foundation when installed. For a successful design with geotextiles in general, and thus also for these geosystems in particular, it is necessary to have information concerning the interaction between the geotextile and other materials in the structure. Performance tests are normally required for this purpose but standardisation of these is difficult since they have to be adjusted for local conditions. Nonetheless, it is important to take account of this whenever test results supplied by the manufacturer are used in a design.

CE Marking contains various distinct functions for which a geotextile can be used:

- reinforcement;
- separation;
- filtration;
- drainage.

To assess whether a geotextile is suitable for a particular application, relevant geotextile index properties must be available for individual geotextile types. The relevant geotextile index properties are described in the European Application Standards EN13249 to EN13257 and EN13265 for geotextile applications. The European Application Standard EN13253 "Geotextiles and geotextile-related products – Required characteristics for use in erosion control works" is relevant for geotextile-encapsulated sand elements.

Appendix B contains further information on CE Marking for geotextiles. It covers the different index tests and the durability of geotextiles in more detail. Table 2.8 gives some standard specifications of geotextiles.

Chapter 3

Geotextile bags

Geotextile bags are geotextile containers, normally filled with sand, and with volumes between 0.3 and 10 m³. If the volume of the bags is lower than 0.3 m³, then they are normally referred to as sandbags (see Figure 3.1) and these are not covered in this manual.

3.1 APPLICATION AND GENERAL EXPERIMENTAL DATA

Geotextile bags are used in a variety of temporary and permanent structures. [22, paragraph 5.2.1] The following applications utilize geotextile bags:

- groynes;
- breakwaters;
- artificial reefs;

Figure 3.1 Geotextile bags used as a beach groyne (Saipem, Angola).

- filling of scour holes;
- soil protection;
- emergency dykes.

The limited dimensions of geotextile bags, compared to larger volume geotextile tubes and geotextile containers, tend to make structures utilizing geotextile bags relatively expensive solutions. Filling is quite labour intensive. In [14, in Dutch] it is claimed that the filling and placement of geotextile bags is around 45–75% of the total cost of a geotextile bag structure. Geotextile bags are therefore used where there are specific advantages, such as:

- there is no requirement to use armour stone. The advantage depends on the unavailability of armour stone;
- the structure is accessible by swimmers (e.g. at a beach);
- the geotextile bags can be placed against other structures, such as piles, with limited risk of damage;
- the structure is relatively easy to remove if it becomes redundant or loses its function.

Optimisation of the fill and installation process lowers the cost of geotextile bag structures and this is a key goal for all geotextile bag usage.

In many cases, however, geotextile bags are used for relatively small-scale applications. Partly as a result of this, there has been little systematic research done to generate design rules for these structures. The literature on the research that has been done on geotextile bags is summarised in [22].

Temporary geotextile bag structures do not always have to be covered. For permanent structures it is necessary to prevent ageing of the geotextile skin by UV light and the bags are covered with armour stone or other material. When used at greater water depth (as in scour holes) no cover is needed since the effect of UV light is low under water. However, a cover layer could be useful against vandalism and other mechanical damage when the geotextile bags are used as exposed surface protection.

When establishing the appropriate weight and dimensions of the geotextile bags an important role is played by experience gained from comparable structures in comparable situations. In Appendix B "Geotextile bags" of [14, in Dutch] several examples are given of practical applications with an indication of dimensions and costs. This presents an initial condition. The designer will have to enter the iterative design process; using his initial assumptions of the dimensions, with the formulae presented below used to check whether or not the structure will be stable under the design conditions. If it is not stable, the design has to be modified accordingly, and this is also be the case if the design proves to be over-conservative and further cost savings can be obtained.

3.2   INSTALLATION PROCEDURE

Geotextile bags can be filled in various ways, mechanically or hydraulically, or by hand.

Placement of geotextile bags can be done using a hydraulic crane equipped with a lifting frame or large grab, on land or on water. Land-based placement may be less suitable for dykes that are difficult to access, or have inadequate stability. For large scour holes it is also possible to dump geotextile bags with a side stone dumper or split-bottom barge. In using the latter method, one has to ensure that no sharp edges are present in the barge that could damage the geotextile bags during dumping. Dumping of geotextile bags has also been done using dump-trucks.

3.3 GEOMETRIC DESIGN

In Figure 2.1 a general design chart is given for geotextile-encapsulated sand elements. The first step in the design process is to establish the functional and technical requirements. This is site specific and therefore falls outside the scope of this manual. As already stated in 2.2, it is assumed that the designer is already at the design process stage and has a clear overview of the functional requirements, has a draft design of the entire structure and wants to detail design the geotextile bag component.

The main dimensions of the structure are established first. Then the sizing of the elements and the determination of the construction procedure is carried out.

Figure 3.2 Example of an elongated geotextile bag [14].

Following this, the economic feasibility is evaluated. Where possible, this should be done in consultation with the operator or contractor.

The thickness of a geotextile bag is not clearly defined, with various methods referred to in the literature. Since the element is flexible, the thickness can also change. For rectangular and more or less round geotextile bags a theoretical diameter can be used (analogous to what is normal for rubble) as if the bag is a cube. The thickness of the bag is then the equivalent to the side of the cube, i.e:

$$D_n = \sqrt[3]{V} \tag{3.1}$$

where:

D_n = nominal thickness of the geotextile bag [m];
V = volume of the geotextile bag [m^3].

For more elongated bags (see Figure 3.2) the above formula would result in an excessive thickness. For well-filled elongated bags the nominal thickness can be better derived using the cross-sectional area as a basis:

$$D_n = \sqrt{A} \tag{3.2}$$

where:

A = cross-sectional area of the geotextile bag perpendicular to the longest axis [m^2].

3.4   FAILURE MODES AND SAFETY CONSIDERATIONS

When designing a structure with geotextile bags, the following potential failure modes must be taken into account:

- rupture of the geotextile due to impact on the bed;
- insufficient seam strength;
- instability of the geotextile bags through wave attack;
- instability of the geotextile bags through water current along the structure;
- instability of the geotextile bags through water current over the structure;
- instability of the geotextile bags due to wave attack;
- foundation failure and deformation;
- liquefaction of the geotextile bag fill.

In the final design phase a safety factor of 1.1 to 1.2 is recommended for geotextile bags in relation to the design value.

During the construction phase, higher factors of safety may be recommended for specific failure mechanisms, especially when the safety of personnel is involved. The dynamic effects associated with filling, loading and installation of the geotextile bags may justify a safety factor as great as 5 for installation effects.

The following section deals with the design formulae for these potential failure modes.

3.5   DESIGN ASPECTS

Once the overall dimensions and the planned construction procedure of the structure have been established, and the size of the geotextile bags has been determined, the detailed design can be carried out. The structure is assessed for each of the potential failure modes shown in the design chart, including the required tensile strength of the geotextile and the stability requirements for waves and currents. If the stability requirements are not fulfilled, a larger, heavier geotextile bag must be selected or the geometry of the structure changed. If the required geotextile tensile strength is too high, a smaller geotextile bag or different operating method can be selected. The following sections investigate the various components of the design cycle in more detail.

3.5.1   Material choice

In the Netherlands geotextile bags of woven polypropylene are normally used. These materials have a greater tensile strength and stiffness than nonwoven geotextiles. A greater tensile stiffness can be an advantage because the bags hold their shape better and do not deform as readily. The disadvantage is that at relatively low strains high localised stresses may cause rupture of the bags. If maintenance of shape is important, polyester geotextile materials may also be used, however, polyester is a smooth material and therefore the bags may move over each other due to the lower surface friction characteristics. Geotextile bags also need to be robust enough to survive impact on the bottom during any dumping operation, see 6.5.3.

3.5.2   Required tensile strength

In addition to filling, the method of placement (by positioning or dumping) governs the required tensile strength of the geotextile used. If this is done utilizing lifting eyes incorporated into the geotextile bags, the whole bag weight will be transferred via the lifting eyes to the geotextile. As the calculation example in this chapter suggests (see 3.7), this approach requires good quality seams and a specific minimum tensile strength of the geotextile.

It is also possible to design and hold the geotextile bag during the filling and installation operation so that it is always fully supported within a mould (see Figure 3.2). In this case the required tensile strength is not directly governed by the weight of the bag.

If a geotextile bag is dumped under water, it must be robust enough not to be damaged when it impacts the bottom. In [5] a calculation method has been developed for geotextile containers, that can also be applied to geotextile bags, where first the fall velocity is calculated and then the corresponding impact stresses. A combination of the formulae gives the required tensile strength of the geotextile (using a safety factor 1):

$$T = \sqrt{2 \cdot \frac{D_n \cdot V}{b \cdot S} \cdot \frac{J}{C_d} \cdot \left(\frac{\rho - \rho_w}{\rho_w} \right) \cdot \rho \cdot g}$$ (3.3)

where:

T = tensile load in the geotextile [N/m];
b = width of the geotextile bag [m];
S = circumference of the geotextile bag (perpendicular to the longest axis) [m];
J = tensile stiffness of the geotextile, see formula 2.3 [N/m];
C_d = drag coefficient (for which 1 may be assumed) [–];
ρ = density of the geotextile-encapsulated sand element [kg/m³];
ρ_w = density of water (1,000 for freshwater 1,030 for saltwater) [kg/m³];
g = acceleration due to gravity (= 9.8) [m/s²];
V = volume of the geotextile bag [m³].

This formula applies when the maximum velocity is reached in water. This is the case for dumping when the water depth is more than 10 m. However, if the bags fall partly through air (for example when dumped from a side stone barge), they quickly acquire speed and can achieve terminal velocity at smaller water depths. Several factors that could lead to a lower required tensile strength, such as the cushioning effect of the sand-fill are not included here. Consequently, the overall factor of safety (γ) of 1.0 can be applied.

In chapter 6 a calculation method is given in which the cushioning effect of the sand-fill is also taken into account. This is applied to geotextile containers but can, in principle, also be used for geotextile bags. Because the shape of geotextile bags is different (relatively rougher) compared to geotextile containers, the method described above is recommended for geotextile bags.

3.5.3 Stability in waves

The stability under wave attack of structural units (blocks, armour stone, gravel, etc.) used in hydraulic engineering is commonly shown in terms of the ratio of wave height to the weight of the structural units. The stability criterion is often a function of the breaker parameter:

$$\frac{H_s}{\Delta_t D_k} = f(\xi)$$ (3.4)

where:

$$\Delta_t = (1 - n) \cdot \frac{\rho_s - \rho_w}{\rho_w}^1$$ (3.5)

---

[1] This formula is different to what is normally used in conventional hydraulic structures. The reason is that the stability of geotextile-encapsulated sand elements as a whole is of importance, and not the stability of the individual grains in the element. The weight of the elements depends on the porosity of the system and thus the term $(1 - n)$ is included in the formula.

For sand-filled systems, $\Delta_t = 0.9$ to 1.0.

$$\xi = \frac{\tan \alpha}{\sqrt{H_s / L_0}} \quad \text{(the breaker parameter)} \tag{3.6}$$

$$L_0 = \frac{g T_p^2}{2\pi} \quad \text{(the wave length in deep water)} \tag{3.7}$$

where:

H_s = significant wave height [m];
D_k = effective thickness of the geotextile-encapsulated sand element [m];
ξ = breaker parameter [−];
Δ_t = relative density of the structural elements [−];
n = porosity of the fill material of the structural elements [−];
α = slope [deg];
L_0 = wave length in deep water [m];
T_p = peak period [s];
ρ_s = density of the fill material which is approximately 2,650 [kg/m³];
ρ_w = density of water (1,000 for freshwater 1,030 for saltwater) [kg/m³].

The effective thickness of the structural elements is partly dependent on the manner in which the units are installed. In Figure 3.3 the two most common installation geometries are shown (I and II).

The situation shown in the diagram concerns a horizontal placement with approximately 50% overlap between adjacent geotextile bags while the bottom diagram shows placement where the adjacent geotextile bags are placed against each other on the slope, comparable with a stone revetment. In the definition of the effective thickness D_k, L is the length of the geotextile bag, with the long side perpendicular to the revetment axis and α the slope of the structure.

According to research, it has been shown [22] that the stability of a structure with geotextile bags depends on the slope angle, the fill material used the height of the slope and the nature of the overlap between the bags. No theory is currently available to deal with all these factors, so a safe approach based on different model tests for a slope of 1:3 and less is presented here. For this slope:

$$\frac{H_s}{\Delta_t D_k} \leq 1.4 \quad \text{(for irregular waves, slope 1:3 or less)} \tag{3.8}$$

For steeper slopes, testing has been carried out on field scale in the Große Wellen Kanal, Hannover, Germany, with bags of 0.15 m³ volume. The results of these tests yield the following stability criterion:

$$\frac{H_s}{\Delta_t D_k} \leq \frac{2.75}{\sqrt{\xi}} \quad \text{(slope 1:3 or steeper)} \tag{3.9}$$

Installation geometry I

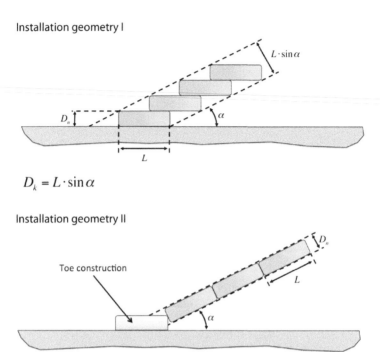

$$D_k = L \cdot \sin \alpha$$

Installation geometry II

$$D_k = D_n$$

Figure 3.3 Definition of effective thickness for installation geometry I and II.

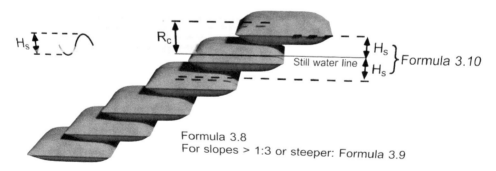

Figure 3.4 Application areas for stability formulae.

This last value (2.75) is close to the value (2.5) mentioned by Pilarczyk [22].

Further, the tests have revealed that the stability of the uppermost elements is less than that of the lower-lying elements. In case the uppermost element is placed between one significant wave height (H_s) above and below the still water line the following relationship applies (see also Figure 3.4):

$$\frac{H_s}{\Delta_t D_k} \leq 0.79 + 0.09 \frac{R_c}{H_s} \quad \text{(for uppermost elements)} \tag{3.10}$$

where:

R_c is the distance between the crest and the still water line, but is zero when the crest is below the still water line.

3.5.4   Stability when subject to longitudinal current flows

To determine the stability of geotextile-encapsulated sand elements when subject to longitudinal current flows (current flow parallel to the structure, see Figure 3.5), as in the application for canals and rivers, use can be made of the Pilarczyk relationship [23], based on a fully protected foundation of sand:

$$\Delta_t D_k \geq 0.035 \cdot \frac{\Phi K_T K_h u_{cr}^2}{\Psi K_s 2g} \tag{3.11}$$

where:

D_k = effective thickness of the geotextile-encapsulated sand element [m];
Δ_t = relative density of the geotextile-encapsulated sand element –
    see formula 3.5 [–];
u_{cr} = critical horizontal flow velocity along the surface of the structure [m/s];
Φ = stability parameter, depending on the application [–];
Ψ = Shields parameter [–];
K_T = turbulence factor [–];
K_h = factor related to the depth [–];
K_s = factor related to the slope angle [–].

For the stability parameter Φ the following values apply:

* for continuous top layer: $\Phi = 1.0$;
* for edges: $\Phi = 1.5$.

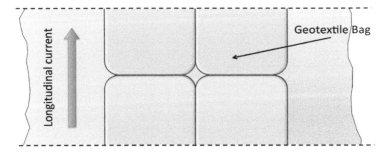

Figure 3.5 View from above of longitudinal current flow in a geotextile bags structure.

Geotextile bags must be designed to account for the stability parameter at the edges ($\Phi = 1.5$) due to the higher forces at the edges.

The Shields parameter depends on the type of element:

- for small geotextile bags (<0.3 m³): $\Psi = 0.035$;
- for larger geotextile bags: $\Psi = 0.05$.

The turbulence factor accounts for the extent of turbulence in the current. In Table 3.1 several values are given.

Using the depth factor (K_h), the depth-average flow velocity is translated into a flow velocity at any depth (h) along the structure:

$$K_h = \frac{2}{\left(\log\left(\dfrac{12h}{k_r}\right)\right)^2} \tag{3.12}$$

where:

h = water depth [m];
k_r = equivalent roughness according to Nikuradse [48] [m].

For the equivalent roughness k_r, an initial estimate of the effective thickness (D_k) of the geotextile bags can be used. It should be noted that formula (3.12) applies to a fully developed current profile. If this is not the case, then the following relationship should apply:

$$K_h = \left(\frac{h}{k_r}\right)^{-0.2} \quad \text{undeveloped current profile} \tag{3.13}$$

$$K_h \approx 1.0 \quad \text{for a very rough current } (h/k_\gamma < 5) \tag{3.14}$$

The slope factor K_s is a function of the influence of the angle of shearing resistance between the geotextile bag and the subsoil:

$$K_s = \sqrt{1 - \left(\frac{\sin \alpha}{\sin \delta}\right)^2} \tag{3.15}$$

Table 3.1 Turbulence factor K_T for various hydraulic conditions [22].

Condition	K_T [–]
Normal turbulence in rivers	1.0
Higher turbulence: river bends	1.5
Turbulence at groynes	2.0
Strong turbulence: hydraulic jumps, sharp bends, local disruptions	2.0
Turbulence as the result of propeller jets and other water jets	3.0–4.0

where:

α = slope angle of the structure [deg];

δ = friction angle between the geotextile bag surface and the subsoil [deg].

Various tests have been performed to determine the friction angle between geotextiles and different subsoils. The results show a large range of friction angles: $\delta = 20°–40°$. For applications under water, or where a subsoil is comprised of wet clay, it is recommended to use a friction angle of $\delta = 20°$. If the calculations result in low stability under longitudinal currents due to a low surface friction angle between the bags and the subsoil slope, it is recommended that actual testing be performed to determine the appropriate geotextile bag/subsoil friction angle. Additionally, a toe structure can be used to prevent the lowest geotextile bags sliding from the slope.

Another aspect that has to be considered when using formula (3.11) is the possibility of sand movement within the geotextile bag. With a continuous external water flow the geotextile bag may be internally stable, but if the sand in the geotextile bag moves, this could result in deformation of the bag and lead to bag instability. This phenomenon was first reported in 1968 [30] in one of the earliest studies into geotextile-encapsulated sand elements. This early small-scale study into the stability under current flows showed that at a flow velocity of more than 1.5 m/s, internal sand movement occurred in the geotextile bag. Moreover, geotextile bags placed on a slope as low as 1:8 would become unstable at a flow velocity of 2.5 m/s because of internal sand movement irrespective of the size of the geotextile bags. Recent research on geotubes showed that also the degree of filling has an influence [41], [33].

3.5.5   Stability when subject to overtopping currents

For some structures, such as soil protection dykes, groynes and underwater dams, the stability when subject to current attack must be checked. The following dimensionless relationship, formula (3.16), between the external current acting on the structure and the resistance (weight) of the geotextile bags can be used.

$$\frac{u_{cr}}{\sqrt{g\Delta_t D_k}} \leq F \tag{3.16}$$

where:

u_{cr} = maximum allowable flow velocity over the crest of the structure [m/s];

F = stability factor [–].

Various values of the stability factor F have been documented:

- $F = 1.2$ [22];
- $F = 0.5 – 1.0$ [9, in Dutch];
- $F = 0.9 – 1.8$ [19].

It is recommended to use a value of $F = 0.9$ for either where an external current is perpendicular to the axis of the structure or where geotextile bags are placed in mounds.

3.5.6 Stability of geotextile bags placed in mounds

The stability of a mound structure of geotextile bags is dependent on the following potential failure mechanisms:

- stability of the mound subject to wave attack;
- shear failure of the subsoil;
- liquefaction of the subsoil.

Stability of the mound under wave attack

During wave attack, if geotextile bags are formed into a mound, a pressure difference may occur between the water pressure within the mound and the external water pressure, and this may cause shearing failures in part of the mound and thus pose a threat to stability. For geotextile containers this phenomenon has been studied in small-scale model tests such as those reported in [6]. In section 6.5.7 the stability of a mound subject to wave attack is discussed in more detail.

Shear failure of the subsoil

If geotextile bags are used for a slope revetment, they must also be checked for potential shearing of the subsoil. A substantial part of this section is extracted from [31, in Dutch].

A wave load on the top layer of a slope penetrates into the subsoil. The load is dampened and delayed due to the water in the porous space being elastically compressible. This creates fluctuating water movements and oscillations of hydrostatic pressure in the subsoil, and thus corresponding oscillations of effective stress. This phenomenon is called elastic recovery and can result in shearing of the subsoil. The stability of the subsoil can be endangered if the elastic recovery causes the effective stress to reduce to the extent that there is insufficient shearing resistance in the subsoil to prevent failure.

The criterion for shearing resistance of the subsoil is considered to be fully 1-dimensional. This criterion is considered to be conservative because extra support can be provided by subsoil elements below the element with the highest loading. The criterion is:

$$\Delta\Phi_{max} \leq (\Delta_t D_k + \Delta_s z) \cdot \left(\cos\alpha - \frac{\sin\alpha}{\tan\delta} \right) \qquad (3.17)$$

where:

$\Delta\Phi_{max}$ = maximum head difference in the level of rise under the top layer of geotextile bags [m];

Δ_t = relative density of the filled geotextile bags, see formula (3.5) [–];

Δ_s = relative density of the subsoil $((\rho_{subsoil} - \rho_w)/\rho_w)$ [–];

z = depth of the subsoil [m].

Design charts have been made of the subsoil shearing criterion for the most common situations in Appendix C. These show the maximum allowable wave height versus the thickness of the top layer for different slope angles. The design charts apply where geotextile bags are located directly on a sandy subsoil, and are given for two wave crests ($s_{op} = 0.03$ and $s_{op} = 0.05$). The wave crest steepness can be determined by:

$$s_{op} = \frac{H_s}{L_0} = \frac{2\pi H_s}{gT_p^2} \qquad (3.18)$$

The design charts in Appendix C list an equivalent thickness of the top layer. If the system is placed directly on sand, the equivalent thickness is equal to the real thickness of the top layer, i.e. $D_{eq} = D_k$. If the system is placed on a granular filter layer, then the weight of the filter layer has a positive effect on the stability of the subsoil. This effect can be expressed in terms of an equivalent thickness, calculated by:

$$D_{eq} = D_k + \frac{\Delta_f}{\Delta_t} \cdot b_f \qquad (3.19)$$

where:
D_{eq} = equivalent thickness of geotextile bags [m];
Δ_f = relative density of the granular filter layer [–];
b_f = thickness of the granular filter layer [m].

Liquefaction

A cyclic load causes a sand layer to compress, which reduces the pore space. The water in the pores pressurises and flows out. Initially, excess hydrostatic pressures occur, causing the contact pressure between the sand grains to decrease and consequently, the resistance to shearing is lowered. In the extreme condition, the excess hydrostatic pressures may be so large that the contact pressure between the sand granules reduces to zero. This condition is known as liquefaction.

Liquefaction can occur during earthquakes as well as during wave attack on slopes. With regard to the latter occurrence, the following design rules apply for structures with reasonably well-compacted subsoil:

- For a top layer on sand there is no danger of liquefaction if the slope:
  - is less than or equal to 1V:3H, or;
  - is less than 1V:2H and the significant wave height H_s is less than 2 m, or;
  - is less than 1V:2H and the subsoil is well-compacted;
- For a top layer on clay there is no danger of liquefaction;
- For a top layer on a granular filter there is generally no danger of liquefaction.

3.6   CONSTRUCTION ASPECTS

From the previous sections, a number of construction aspects warrant attention:

- Filling geotextile bags can be a labour-intensive process, which has consequences with regard to the materials used and the number of people employed.

- Specific methods exist in which the geotextile bag is constantly supported during filling and placement;
- The use of a side stone dumper or split-bottom barge should contain no sharp edges that may damage the geotextile bags.

3.7 CALCULATION EXAMPLE

Given:

A bank protection has to be placed in a river with normal turbulence and a depth of 3.4 metres. The structure will be constructed using geotextile bags. The water current profile is regarded as not being fully developed (whether the current profile is fully developed or not is often difficult to determine). A safe assumption has been made and the following data are known (*see also Figure 3.6*):

- significant wave height: $H_s = 0.85$ m;
- peak period: $T_p = 4.75$ s;
- porosity of the geotextile bag fill material: $n = 0.45$ [–];
- density of sand: $\rho_s = 2,650$ kg/m³;
- density of water: $\rho_w = 1,000$ kg/m³;
- surface friction angle between the geotextile bags and the subsoil: $\delta = 20°$;
- slope angle: $\alpha = 1V:3H$ (18.43°);
- acceleration due to gravity: $g = 9.81$ m/s²;
- distance from crest to still water line: $R_c = 1.1$ m;
- maximum water flow velocity along the slope: $u_{cr} = 1.5$ m/s;
- water depth: $h = 3.4$ m;
- the top layer is continuous.

The geotextile bags have the following characteristics:

- width: $b = 2.5$ m;
- length: $l = 1.5$ m;

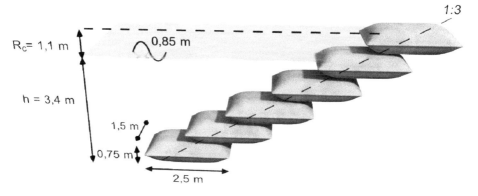

Figure 3.6 Description of the various parameters used in this example.

- height: $h = 0.75$ m;
- geotextile material: woven polypropylene (PP);
- geotextile tensile strength: $T_m = 120$ kN/m;
- geotextile strain at maximum tensile strength: $\varepsilon_m = 0.14$.

Required solution

Check the prescribed geotextile bags for stability in waves and current. Also determine if the tensile strength of the geotextile bags is sufficient if they are to be dumped from a barge.

Calculation method

The design parameters are determined using formulae (3.5), (3.7) and (3.6):

$$\Delta_t = \left(1 - n\right) \cdot \frac{\rho_s - \rho_w}{\rho_w} = \left(1 - 0.45\right) \cdot \frac{2650 - 1000}{1000} = 0.91$$

$$L_0 = \frac{g \cdot T_p^2}{2 \cdot \pi} = \frac{9.81 \cdot 4.75^2}{2 \cdot \pi} = 35.23 \text{ m}$$

$$\xi = \frac{\tan \alpha}{\sqrt{\dfrac{H_s}{L_0}}} = \frac{\tan 18.43}{\sqrt{\dfrac{0.85}{35.23}}} = 2.15$$

The effective thickness of the geotextile bag can be determined using formula (3.4):

$$D_k = L \cdot \sin\alpha = 2.5 \cdot \sin(18.43) = 0.79 \text{ m}$$

Stability subject to waves

The next step is to check the stability subject to waves. The slope has an incline of 1V:3H. Using formula (3.8):

$$\frac{H_s}{\Delta_t D_k} \le 1.4$$

$$\frac{0.85}{0.91 \cdot 0.79} = 1.18 \ \Rightarrow \ \text{complies}$$

For the geotextile bags that will be installed between SWL $\pm H_s$ the equilibrium is determined using formula (3.10):

$$\frac{H_s}{\Delta_t D_k} \le 0.79 + 0.09 \frac{R_c}{H_s}$$

$$= 0.79 + 0.09 \frac{1.1}{0.85} = 0.91 \Rightarrow \text{does not comply}$$

The minimum required effective thickness for these geotextile bags is determined using:

$$\frac{\dfrac{H_s}{\Delta_t}}{0.79+0.09\dfrac{R_c}{H_s}} = \frac{\dfrac{0.85}{0.91}}{0.79+0.09\dfrac{1.1}{0.85}} = 1.03 \text{ m}$$

Therefore, in terms of the minimum required length of these geotextile bags, the following relationship applies:

$$L \geq \frac{D_k}{\sin\alpha} = \frac{1.03}{\sin 18.43} = 3.26 \text{ m}$$

Stability subject to longitudinal water current

To check stability subject to longitudinal water current, K_s is first calculated according to formula (3.15):

$$K_s = \sqrt{1-\left(\frac{\sin\alpha}{\sin\delta}\right)^2} = \sqrt{1-\left(\frac{\sin 18.43}{\sin 20}\right)^2} = 0.38$$

Using a geotextile bag, the equivalent roughness according to Nikuradse is:

$$k_r = D_k = 0.79 \text{ m}$$

The depth parameter K_h may be determined using formula (3.13):

$$K_h = \left(\frac{h}{k_r}\right)^{-0.2} = \left(\frac{3.4}{0.79}\right)^{-0.2} = 0.75$$

Now, the required thickness can be calculated using formula (3.11):

$$\Delta_t D_k \geq 0.035 \cdot \frac{\Phi K_T K_h u_{cr}^2}{\Psi K_s 2g} = 0.035 \cdot \frac{1.5\cdot 2\cdot 0.75\cdot 1.5^2}{0.05\cdot 0.38\cdot 2\cdot 9.81} = 0.48 \text{ m}$$

$$D_k \geq \frac{0.48}{\Delta_t} = \frac{0.48}{0.91} = 0.53 \text{ m} \Rightarrow \text{complies}$$

Stability subject to current over the structure

In the case of well-designed bank protection there is no current over the structure.

Required geotextile tensile strength

The required geotextile tensile strength when dumping the geotextile bags can be determined using formula (3.3):

$$T = \sqrt{2 \cdot \frac{D_n \cdot V}{b \cdot S} \cdot \frac{J}{C_d} \cdot \left(\frac{\rho - \rho_w}{\rho_w} \right) \cdot \rho \cdot g}$$

The geotextile bags are produced from woven polypropylene with a tensile strength of 120 kN/m at a corresponding strain of 14%. The tensile stiffness modulus J of the geotextile can be determined using formula (2.2):

$$J = \frac{T_m}{\varepsilon_m} = \frac{120}{0.14} = 857 \text{ kN/m}$$

The tensile load in the geotextile can now be determined by (J in formula 3.3 has to be presented in N/m):

$$T = \sqrt{2 \cdot \frac{0.75 \cdot 1.875}{2.5 \cdot 6.5} \cdot \frac{857 \cdot 10^3}{1} \cdot \left(\frac{1908 - 1000}{1000} \right) \cdot 1908 \cdot 9.81}$$

$T = 50$ kN/m

This is the theoretically calculated tensile load that any seam in the geotextile bag must be able to resist. If it is assumed that the seams achieve a strength of 70% of the strength of the geotextile, the required tensile strength of the geotextile is then:

$$T_{geotextile} = \frac{T_{seam}}{0.7} = \frac{50}{0.7} = 72 \text{ kN/m}$$

An overall safety factor $\gamma = 1.0$ may be applied since the formula for determining the required tensile strength already has a number of safe assumptions:

$$\frac{T_m}{T_{geotextile}} = \frac{120}{72} = 1.67 \Rightarrow \text{complies}$$

Sand movement within the geotextile bag

The water flow is less than or equal to 1.5 m/s, so there is little likelihood of sand movement within the geotextile bag.

Stability of the mound under wave attack

For geotextile containers the stability of the mound under wave attack has been studied using small-scale model tests. Reference should be made to the calculation example for geotextile containers in section 6.7.

Shear failure of the subsoil

As far as shearing of the subsoil is concerned, the design charts in Appendix C show the maximum allowable wave heights at a $D_{eq} = D_k = 0.79$ m at a slope of 1V:3H. The wave crest steepness $s_{op} = 0.03$ has been used (Figure C.1 in Appendix C): $H_s < 1.2$ m. In this calculation example, $H_s = 0.85$ m was the design significant wave height and consequently, the stability requirement is fulfilled.

Liquefaction

No danger of liquefaction exists in view of the slope being less than 1V:3H.

Chapter 4

Geotextile mattresses

Geotextile mattresses comprise two interconnected layers of geotextile where the space between is filled with sand and, in special cases, concrete. Cells or tubes form compartments within the mattress, which facilitates an even distribution of the fill material in the geotextile mattress and maintains its shape and combats movement of the fill material during use.

Geotextile mattresses are manufactured from woven geotextile materials or from combinations of woven and nonwoven geotextile materials. The choice of geotextile depends on the hydraulic, mechanical and environmental requirements of the method of construction.

4.1 APPLICATION AREAS AND GENERAL EXPERIMENTAL DATA

Geotextile mattresses are used as bank or bed protection along rivers and canals, often in countries where there is no natural bank protection materials (blocks or rock) available and/or where there is a lack of technical personnel to construct proper granular filters or similar layers. Figure 4.1 shows a typical geotextile mattress used for a bank revetment.

To determine the appropriate dimensions of geotextile mattresses, use is often made of the design approach for sloping revetments using set stone. For geotextile mattresses, the drainage length of the structure may also be expected to have some influence. The drainage length is determined by the relationship between the hydraulic conductivity of the mattress layer and that of the subsoil. With a relatively impermeable subsoil, the mattress is prone to become unstable more quickly; with a relatively permeable subsoil, instability occurs less quickly, but the load on the subsoil is greater.

When geotextile mattresses are used, they are normally placed directly on the subsoil. In some instances they may be supplemented by a sand layer as an extra fill layer.

The appearance of a mattress can be of flat-tubular form or can be waffle-like. Sometimes the tubes or 'sausages' of the waffle-like geotextile mattress are placed horizontally on the slope, but more often the tubes are placed down the direction of the slope (average 1V:2H to 1V:4H) – see section 4.3. Placing the mattresses down the direction of the slope also enables easy filling (by gravity) on site. In this direction, the seams between the tubes in the mattress follow the inclination of the slope. Because

Figure 4.1 Use of a geotextile mattress as bank protection at Nageler Vaart (The Netherlands).

the geotextile mattress in the vicinity of the seams comprises only two geotextile layers and is thus at that location quite thin and permeable, the hydraulic gradient on the filter and the subsoil is therefore relatively high. In such a situation the critical values for the hydraulic gradient and velocity may be exceeded and this may cause local displacement of the base material. By filling the sausages to the maximum volume the seams are under tension and the average thickness of the mattress increases. Proper installation ensures the hydraulic gradient on the filter is reduced and any subsequent displacements minimized.

Maximum sand density inside the mattresses also has the effect that internal sand compaction is maximized and movement by waves and currents is minimized. Mattresses always need to be anchored on the upper side to prevent sliding down the slope. This is done with the use of an anchor trench.

4.2 INSTALLATION PROCEDURE

Geotextile mattresses are filled at ground level or in-situ on the slope. Filling can be carried out pneumatically, mechanically or hydraulically. These methods are further described in [22, section 5.2.2].

Most commonly, mattresses are filled hydraulically. By hydraulic filling the highest possible filling density can be achieved. With this method the fill pressure of the sand-water mixture plays a role in achieving maximum filling density. During filling, the pressure in the mattress must be prevented from rising too high, especially if the geotextile silts up during the filling process.

Preferably, geotextile mattresses should be filled in-situ, which minimizes the tensile loads generated in the geotextile. In case in-situ filling is not possible, the geotextile mattresses can be filled at ground level and transferred by a pontoon and installed in place using a clamping system, or be lifted into place with a crane. In both cases significant tensile loads will be generated in the geotextile, in which case the geotextile used must take into account these additional installation tensile loads. It should be noted that a well-filled mattress is important in order to prevent the loss of sand in the mattress when hoisted.

The method of construction chosen is determined by the size of the mattress, the available space in which to work, accessibility from the bank and from the water, economic feasibility and the experience of the installer.

4.3 GEOMETRIC DESIGN

In Figure 2.1 a general design procedure is given for geotextile-encapsulated sand elements. The first step in the design procedure is to establish the basic functional and technical requirements. This is an area that falls outside the scope of this manual. As already discussed (see 2.2), it is assumed that the designer is already at the design stage and has a clear understanding of the functional requirements, has a draft design of the overall structure and wants to go through the detailed stage.

The overall dimensions of the structure are established first. This is followed by the size of the elements and the construction of the structure based on experiential data, constructability, economic feasibility and application area. Where possible, this is done in consultation with the installer.

The common width of standard geotextile roll materials is 4.50 to 5.30 m. By sewing more strips to each other larger continuous sheets can be made, thus reducing the amount of overlap. The strip width is limited by the amount of space available at the construction site and by handling and placement issues. If the mattress is filled in-situ, there are less limitations regarding dimensions.

There are several different types of geotextile mattresses [22]. The most common are listed below (see also Figure 4.2):

*   *The woven mattress.* This involves two geotextiles, one above the other, woven together on a loom where the upper and lower geotextiles are woven together at regular intervals. This process requires special production equipment.

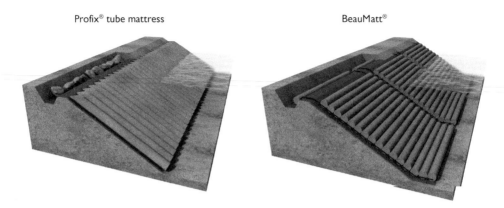

Profix® tube mattress   BeauMatt®

Figure 4.2 Profix® tube mattress and the BeauMatt® mattress.

- *The tube mattress.* This is the most common type of mattress. The geotextile layers are sewn together lengthwise so that parallel tubes are formed that can be filled with sand. Since the 'flat' tubes are inflated when filled, the tube shrinks cross-wise by approximately 20–25%. Depending on the distance between seams, different diameter tubes can be formed.
- *The shallow spherical mattress.* This mattress undergoes the same production process as the tube mattress but less geotextile is used for the lower mattress enabling the tubes to become a semi-cylindrical shape after filling, and where the mattress does not contract during filling. The mattresses can be placed such that the tubes lie in the direction down the slope or horizontal to it.

For the stability calculations, a distinction must be made between the thickness of the tubes in the geotextile mattress and the average thickness of the geotextile mattress itself. In general:

$$\frac{D_k}{D_d} = 0.6 \text{ to } 0.8 \quad (90\% \text{ to } 95\% \text{ filled}) \tag{4.1}$$

where:

D_k = effective thickness of the geotextile mattress [m];
D_d = maximum diameter of the mattress tubes [m].

A frequently used (average) mattress thickness is 0.20 m.

It should be noted that many similarities exist between the design process for geotextile mattresses and that for geotextile bags. This does not mean that there is no difference between the designs using the two types of encapsulated sand elements. For example, the potential failure mechanisms for the overall structure will differ considerably and thus require different design approaches.

4.4   POTENTIAL FAILURE MECHANISMS AND SAFETY CONSIDERATIONS

For the design of a structure using geotextile mattresses assessment of the following potential failure mechanisms must be made:

- Rupture of the geotextile during handling and placement of the geotextile mattress;
- Rupture of the geotextile as a result of anchoring – the additional tensile loads have not been taken into account;
- Rupture of the geotextile through insufficient seam strength;
- Instability of the geotextile mattress due to wave attack;
- Instability of the geotextile mattress due to water flows parallel to the structure;
- Shear failure of the subsoil;
- Liquefaction of the subsoil;
- Vandalism and recreational damage.

For the design or construction phase, normally a safety factor of 1.1 to 1.2 is applied to the thickness of geotextile mattresses in relation to the design value. During construction, extra attention must be paid to safety when the pulling and hoisting of mattresses is required.

The following section deals with the design formulae associated with the above listed potential failure mechanisms.

4.5   DESIGN ASPECTS

Once the overall dimensions of the structure, the construction procedure and the size of the elements have been determined, the detailed design is carried out. The structure is assessed in respect of the stages shown in the design procedure, including the required tensile strength for the geotextile and the stability requirements for waves and water flows. If the stability requirements are not fulfilled, a heavier geotextile mattress must be selected, having a larger volume and higher geotextile tensile strength. If the required tensile strength of the geotextile is not cost-effective, a lighter and thinner geotextile mattress and/or different operating method may be selected. The geotextile mattress is then checked against the stability requirements for waves and water flows, and for geotechnical stability. Detailing at the edges of the geotextile mattress is important to ensure no localised instability can occur and in many instances this can be the critical part of the design.

4.5.1   Material choice

Most mattresses are made of polypropylene geotextile materials, using heavier geotextiles to limit the effects of UV radiation and mechanical damage from vandalism, etc. This is why the top geotextile layer commonly consists of a combination of geotextile materials (to resist long term UV exposure and provide resistance to vandalism, etc.).

Where a filled geotextile mattress has to be hoisted or dragged into place without causing damage a high strength bottom geotextile layer is required.

4.5.2   Required tensile strength

The required tensile strength of a geotextile mattress is largely governed by the method of construction. During filling, hoisting and pulling, and placement, major drag and tensile loads can be exerted on the geotextile mattress. In addition to the strength of the geotextile and the seams, anchoring at the edges is also an important aspect.

Hoisting

Where a clamping system is used, the maximum tensile load in the geotextile mattress during the hoisting process (see Figure 4.3) can be determined as:

$$T = \frac{l \cdot b_m \cdot D_k}{b_g} \cdot \rho \cdot g \qquad\qquad (4.2)$$

where:

T = tensile load per unit width in the geotextiles in the mattress [N/m];
l = length of the geotextile mattress [m];
b_m = width of the geotextile mattress [m];
b_g = width of the geotextile [m];
D_k = effective thickness of the geotextile mattress [m];
ρ = density of the geotextile mattress [kg/m³];
g = acceleration due to gravity [m/s²].

Figure 4.3 Tensile loads in geotextile mattress during hoisting.

Anchoring

Since the geotextile mattress cannot be supported by a toe provision, the mattress must be anchored adequately on the upper side. For the strength and stability of these anchoring structures, general calculation rules have been derived. For the maximum tensile load on the anchoring in still water, the following applies (see also Figure 4.4):

$$T = (\sin\alpha - \cos\alpha \cdot \tan\delta) \cdot (L_{bw} \cdot \rho \cdot g + L_{ow}(\rho - \rho_w) \cdot g) \cdot D_k \qquad (4.3)$$

where:

T	=	tensile load per unit width in the geotextile mattress [N/m];
L_{bw}	=	length of the geotextile mattress above water [m];
L_{ow}	=	length of the geotextile mattress under water [m];
α	=	slope angle [deg];
δ	=	surface friction angle between the unanchored geotextile mattress and the subsoil [deg].

If the bottom geotextile layer is fairly rough, the friction angle between the underside of the geotextile and the bank is of the same order as the bank material itself. For bank materials consisting of sand and other granular materials, $\delta = 30°–35°$. If the frictional behaviour of the bottom geotextile layer cannot be established (e.g. as in the case of geotextile bags), it is advisable to adopt a conservative value for the friction angle, ($\delta = 20°$ for a slope of sand).

During wave attack, when the waves recede, the mattress may make reduced contact with the subsoil due to the water flowing out and this can result in a reduction in stability. The location where the mattress makes reduced contact can only be calculated when there is knowledge of the hydraulic conductivity of both the mattress and the subsoil. This manual has opted for a conservative approach to the assessment of

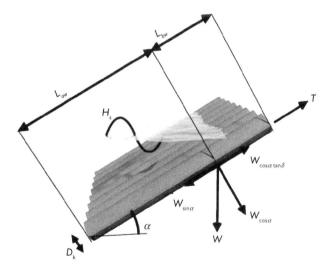

Figure 4.4 Tensile loads in the geotextile mattress as a result of anchoring. L_{ow} and L_{bw} are the lengths along the mattress.

the tensile load in the mattress where it is assumed that over one wave height there is no loss of contact under the water line between the mattress and the subsoil. The formula is then:

$$T = (\sin\alpha - \cos\alpha \cdot \tan\delta) \cdot L_{bw} \cdot D_k \cdot \rho \cdot g$$
$$+ \left(\sin\alpha - \left(1 - \frac{H_s}{L_{ow}}\right) \cdot \cos\alpha \cdot \tan\delta\right) \cdot L_{ow} \cdot D_k \cdot (\rho - \rho_w) \cdot g \qquad (4.4)$$

With formula (4.4) it is possible to calculate a negative value for T, which means that the geotextile mattress remains on the slope under its own weight, with no tensile load occurring at the anchoring point.

For equilibrium along the slope:

$$T = W \cdot \sin\alpha - W \cdot \cos\alpha \cdot \tan\delta \qquad (4.5)$$

In the event of unforeseen conditions, e.g. flow shearing along the underside of the geotextile mattress, the loads on the anchoring point can be high (up to a maximum of the total weight of the mattress) and be coupled with substantial tensile loads in the geotextile mattress.

In [14, in Dutch] a falling apron is also given as a possible application. In such a structure the subsoil can erode from under the mattress as there is reduced contact between the mattress and the subsoil. When determining the tensile loads in the mattress in this case the friction between the mattress and the subsoil is disregarded. The anchoring must therefore be able to support the weight of any part of the mattress that is in free suspension. Only the component of that part of the mattress where the subsoil does not erode, can contribute to the stability through friction between the subsoil and mattress.

Application phase

External damage to the upper geotextile layer is another possible cause of mechanical damage to the geotextile mattress. This can occur by vandalism or recreational damage, e.g. by impact from a boat. However, the compartmentalisation attributes of the mattress limit the extent of any damage:

* With damage caused by impact from a boat, the deeper part of the geotextile mattress remains filled with sand;
* The use of a combination of materials with a protection layer on the upper geotextile layer strongly limits mechanical damage and hinders vandalism to the upper surface;
* Where the upper mattress surface is damaged, there is still approximately 50% of the strength in the lower mattress surface before the slope becomes exposed.

For the application phase, the actual density of the fill material is important. Wave loads and water flows may cause the fill material inside the mattress to compress further. If the original density is too low, further compression will reduce the mattress thickness and this may cause gaps to occur between the mattress and the subsoil, which may result in local deformation of the subsoil.

4.5.3   Stability in waves

In general, geotextile mattresses should not be used where the significant wave height H_s is greater than 1.0 m. [22, paragraph 5.4.4] gives the following empirical stability ratio:

$$\frac{H_s}{\Delta_t \cdot D_k} \leq \frac{4 \text{ to } 5}{\xi^{2/3}} \quad \text{for } H_s < 1.0 \text{ m} \tag{4.6}$$

where:

H_s = significant wave height [m];
Δ_t = relative density of the geotextile mattress [–];
D_k = effective thickness of the geotextile mattress [m];
ξ = breaker parameter [–].

The breaker parameter is the ratio between the slope angle and the wave steepness, see formula (3.6). For ship induced waves, the wake of a ship has a general wave steepness ($s_{op} = H_s/L_0$) of 0.035 to 0.050. The wave height is mostly less than or equal to 0.5 m and the wave length is 10 to 15 m.

4.5.4   Stability in longitudinal currents

To determine the stability of geotextile mattresses in longitudinal currents, use can be made of the Pilarczyk relationship [23], as is done for geotextile bags:

$$\Delta_t D_k \geq 0.035 \cdot \frac{\Phi \cdot K_T \cdot K_h \cdot u_{cr}^2}{\Psi \cdot K_s \cdot 2 \cdot g} \tag{4.7}$$

where:

u_{cr} = critical flow velocity along the structure [m/s];
Φ = stability parameter [–];
Ψ = Shields parameter [–];
K_T = turbulence factor [–];
K_h = factor related to the depth [–];
K_s = factor related to the slope angle of the bank;
g = acceleration due to gravity [m/s^2].

For the stability parameter the following values apply (as for geotextile bags):

* for continuous top layer: $\Phi = 1.0$;
* for edges: $\Phi = 1.5$.

For geotextile mattresses the Shields parameter (a guide value) is [23]: $\Psi = 0.07$. The turbulence factor describes the extent of turbulence in the water flow, see Table 3.1. Using the depth factor (K_h) the depth-averaged flow velocity is translated into a flow velocity just above the toe of the structure. For the determination of this, see formulae (3.12) to (3.14).

The slope factor K_s is a function of the influence of the angle of shearing resistance between the geotextile mattress and the subsoil:

$$K_s = \sqrt{1 - \left(\frac{\sin \alpha}{\sin \delta}\right)^2} \qquad (4.8)$$

where:
  α = slope of the structure [deg];
  δ = surface friction angle between the geotextile mattress and the subsoil [deg].

In most cases the mattresses are anchored at the top of the slope. If this anchoring is sufficient to prevent sliding $K_s = 1$. Using the formulae above, it is apparent that in normal conditions, geotextile mattresses will not be subject to critical loads under normal current conditions.

4.5.5   Geotechnical stability

Geotextile mattresses are mostly used for bank protection. In this situation they also have to be evaluated for geotechnical stability. In chapter 3 the potential mechanisms of shearing of the subsoil and liquefaction within geotextile bags are described (see 3.5.6). These potential mechanisms may also occur when geotextile mattresses are used. For this the design graphs provide in Appendix C, can be used which have been derived for both geotextile mattresses and geotextile bags. In these graphs the geotechnical calculation assumes that the geotextile mattress is not anchored and that it has to remain stable over its full design life. For geotextile mattresses this is a conservative assumption, while for geotextile bags, without interconnections, this is more realistic. A slope protected with an anchored geotextile mattress will not necessarily suffer a loss of strength if the design graphs shown in Appendix C indicate that the structure is unstable. In this state, strain will occur in the geotextile mattress and cause deformation. Since this may be an undesirable condition and affect serviceability, reference should be made to Appendix C. If the deformations are acceptable, then the structure may be also stable under higher wave heights, although this depends on the extent of the deformations.

4.6   CONSTRUCTION ASPECTS

For geotextile mattresses, the following aspects require attention during construction:

- With the in-situ filling method, mattresses can be sewn together which enables, larger sections to be laid, minimizing the material losses due to overlaps. It is advisable, however, to apply an overlap of approximately 1.5 metres after each 4th or 5th mattress so as to limit the loads on the seams due to possible subsequent settlements;
- During in-situ filling the width of the mattress will decrease in the direction perpendicular to the tube. This may lead to extra tension loads in the mattress depending on the applied filling procedure;

- The mattress must be filled to the maximum fill density;
- The mattress must make maximum contact with the slope surface;
- For pulling or hoisting during installation the dimensions (length and/or width) should be limited to restrict the tensile loads acting on the mattress;
- Overlaps between mattress sections can be made by leaving 1 or 2 tubes of the previous mattress empty (expensive) or equipping the previous mattress with a sewn-on flap;
- Always anchor the mattresses at the upper edge.

4.7 CALCULATION EXAMPLE

Given:

Bank protection has to be installed in a canal with shipping. A structure using geotextile mattresses, as shown in Figure 4.5, is considered.

The following data are known:

- significant wave height: $H_s = 0.50$ m;
- peak period: $T_p = 2.1$ s;
- flow velocity along the slope: $u_{cr} = 1.5$ m/s;
- water depth $h = 3.4$ m;
- slope: $\alpha = 1V{:}2.5H$ (21.8°);
- porosity of sand: $n = 0.40$ [–];
- density of sand: $\rho_s = 2,650$ kg/m³;
- density of water: $\rho_w = 1,000$ kg/m³.

The geotextile mattresses to be used have the following properties:

- tube diameter: $D_d = 0.25$ m;
- average thickness of the mattress: $D_k = 0.20$ m;

Figure 4.5 Definition sketch of the calculation example for geotextile mattresses.

- length of the mattress: $l = 10$ m;
- width of the mattress: $b_g = 5$ m;
- surface friction angle between the geotextile mattress and the subsoil: $\delta = 25°$;
- geotextile: woven polypropylene (PP);
- tensile strength of geotextile: $T_m = 80$ kN/m;
- geotextile strain at maximum tensile strength: $\varepsilon = 0.10$.

Required

Determine whether or not the tensile strength of the geotextile mattress is sufficient if it is hoisted fully to its end position. Also check whether the tensile strength is sufficient during the application phase when the geotextile mattress is anchored. Verify that the geotextile mattresses are stable in waves and currents, and have adequate geotechnical stability.

The density of the geotextile mattress in dry conditions is:

$$\rho_{mat} = (1 - n) \cdot \rho_s = (1 - 0.4) \cdot 2650 = 1590 \text{ kg/m}^3$$

and for a fully saturated geotextile mattress its density will be:

$$\rho_{mat} = (1 - n) \cdot \rho_s + n \cdot \rho_w = (1 - 0.4) \cdot 2650 + (0.4 \cdot 1000) = 1990 \text{ kg/m}^3$$

Required tensile strength

Assuming the geotextile mattress will be filled hydraulically, the maximum tensile load in the geotextile mattress during hoisting is calculated according to formula (4.2):

$$T = \frac{l \cdot b_m \cdot D_k}{b_g} \cdot \rho_{mat} \cdot g = \frac{10 \cdot 4.0 \cdot 0.20}{5.0} \cdot 1990 \cdot 9.81 = 31,200 \text{ N/m} = 31.2 \text{ kN/m}$$

where:
$b_m = 4.00$ m (= approx. 80% of empty geotextile width b_g due to shortening of the mattress during filling).

To determine the maximum tensile load as a result of anchoring, it is assumed that the mattress is 7.5 m under water (L_{ow}) and 2.5 m above water (L_{bw}). This is derived from a slope of 1:2.5 and a water depth of 3.4 m. According to formula (4.4):

$$T = (\sin \alpha - \cos \alpha \cdot \tan \delta) \cdot L_{bw} \cdot D_k \cdot \rho_{mat} \cdot g$$
$$+ \left(\sin \alpha - \left(1 - \frac{H_s}{L_{ow}}\right) \cdot \cos \alpha \cdot \tan \delta \right) \cdot L_{ow} \cdot D_k \cdot (\rho - \rho_w) \cdot g$$

$$T = (\sin(21.8) - \cos(21.8) \cdot \tan(25)) \cdot 2.5 \cdot 0.20 \cdot 1990 \cdot 9.81$$
$$+ \left(\sin(21.8) - \left(1 - \frac{0.50}{7.5}\right) \cdot \cos(21.8) \cdot \tan(25) \right) \cdot 7.5 \cdot 0.20 \cdot (1990 - 1000) \cdot 9.81$$
$$= -1.08 \text{ kN/m}$$

The negative value for T means that the geotextile mattress stays on the slope under its own weight and that under these circumstances no tensile load occurs on the anchoring.

Thus, the maximum tensile load occurs during hoisting. In reality, the upper geotextile will also contribute to the absorption of the tensile load. This is, however, not assumed in the calculation above and the tensile load is absorbed by a seam, whose tensile strength is 60% (this % depends on the type of seam and material used, see section 2) of the tensile strength of the geotextile. The required tensile strength of the geotextile is thus (conservative approach as the influence of surface friction is not taken into account):

$$T_{geotextile} = \frac{T_{seam}}{0.6} = \frac{31.2}{0.6} = 52 \text{ kN/m}$$

The available polypropylene geotextile has a tensile strength of 80 kN/m. For the safety factor this means:

$$\gamma = \frac{T_m}{T_{geotextile}} = \frac{80}{52} = 1.54$$

Stability in waves

In this example H_s is 0.50 m and formula (4.6) can be used to check the stability in waves:

$$\frac{H_s}{\Delta_t \cdot D_k} \leq \frac{4 \text{ to } 5}{\xi^{2/3}}$$

$$\frac{0.5}{0.98 \cdot 0.2} \leq \frac{4 \text{ to } 5}{1.5^{2/3}} \quad \Rightarrow \quad 2.55 \leq 3.05 \text{ to } 3.82 \quad \Rightarrow \quad \text{complies}$$

Stability in longitudinal currents

For the calculation of stability in longitudinal currents the Pilarczyk relationship, formula (3.11), is used:

$$\Delta_t D_k \geq 0.035 \cdot \frac{\Phi \cdot K_T \cdot K_h \cdot u_{cr}^2}{\Psi \cdot K_s \cdot 2 \cdot g}$$

- Stability parameter $\Phi = 1.5$ (at the edges);
- Shields parameter $\Psi = 0.07$;
- $K_T = 1.5$ (see Table 3.1).

Now, from formulae (3.15) and (3.13):

$$K_s = \sqrt{1 - \left(\frac{\sin(21.8)}{\sin(25)}\right)^2} = 0.48$$

$$K_b = \left(\frac{b}{k_r}\right)^{-0.2} = \left(\frac{3.4}{0.20}\right)^{-0.2} = 0.57$$

Thus:

$$\Delta_t \cdot D_k \geq 0.035 \frac{1.5 \cdot 1.5 \cdot 0.57 \cdot 1.5^2}{0.07 \cdot 0.48 \cdot 2 \cdot 9.81} = 0.15 \text{ m}$$

$$D_k \geq \frac{0.15}{0.98} = 0.15$$

Shearing of the subsoil

As far as the shearing of the subsoil is concerned, the design graphs in Appendix C show maximum allowable wave heights at $D_{eq} = D_k = 0.20$ m and at a slope of 1V:2.5H. For a wave steepness of $s_{op} = 0.03$ and $s_{op} = 0.05$, $H_s < 0.3$ m applies.

In this calculation example $H_s = 0.50$ m has been used. The relatively steep slope means there is a risk of a shear failure of the subsoil, but the risk is small because, as mentioned previously, the design graphs are conservative for mattresses. If required, possible alleviation measures would be to lessen the slope angle (to approx. 1V:3H), making the geotextile mattress thicker, or using a toe structure.

Liquefaction

No danger of liquefaction exists in view of the slope being no steeper than 1V:2H and the wave height not higher than 2.0 m (see 3.5.6).

Chapter 5

Geotextile tubes

Geotextile tubes consist of a water permeable, sand-sealed geotextile filled with sand or other granular materials. The tube diameter can vary from 0.5 m to 5.0 m and the length may vary from 25 to 100 m depending on the project where the geotextile tubes will be installed.

5.1 APPLICATIONS AND GENERAL EXPERIMENTAL DATA

Geotextile tubes are often used in coastal areas, functioning as beach groynes, break-waters, dune toe protection, submerged reefs, containment dykes (see Figure 5.1) or core structures. They offer extra protection against flooding during heavy storms and regulate sediment transport. For hydraulic structures, geotextile tubes can also serve to temporarily replace rip-rap, especially in temporary works and locations where there is a lack of stone. Where a structure is required to be permanent, the inner granular layers can be replaced by geotextile tubes, with the outer armour stone remaining. Applications

Figure 5.1 Construction using geotextile tubes (Incheon bridge, South Korea).

in the Netherlands include a scour protection in a river bend improvement in the Waal at Erlecom (1989), the core of containment dykes of the Naviduct at Enkhuizen Krabbersgat (2003) and the core of auxiliary dykes along Enkhuizen–Lelystad (1998). An extensive description of these projects, can be found in [14, in Dutch]. Numerous international examples exist of this technique, for example the Incheon Bridge project [45, 46] and a submerged barrier in Tuscany [47]. In addition, studies have been executed to investigate the feasibility of continuous geotextile tubes [27].

Another application using geotextile tubes involves the dewatering of dredged material to decrease its volume. By selection of the correct geotextile, surplus water can drain out of the tube and leave behind the solidified soil in the geotextile tube. This manual focuses on sand-filled geotextile tubes, so for further reference to the draining of dredged material using geotextile tubes readers should consult Appendix 5.4 from [22].

5.2   INSTALLATION PROCEDURE

5.2.1   General

A geotextile tube is delivered to the construction site on a roll on a steel pipe. The geotextile tube is unrolled in the correct location with the inlet and outlet ports centred vertically on the upper side. Flexible filling ports (made from a suitable geotextile) with a diameter of around 0.5 m can be placed at a distance of approximately 15 m relative to each other along the crest of the geotextile tube. However, if the project desires more filling ports, they can be placed at shorter spacings. The distance over which a tube can be filled from a single filling port depends on the grain size of the fill material used.

Filling of a geotextile tube is accomplished by hydraulically pumping a mixture of sand and water into the tube, see Figure 5.3. The initial amount of sand of the sand-water mixture entering the tube blocks the geotextile pores to a large extent, reduces its permeability, and leads to a pressure build up. This leads to a rounded "mushroom" shape of the geotextile tube. The pumping water drains through the geotextile skin, and when the tube is nearly filled the pumping water is also drained via the filling ports. To obtain an even filling height adjacent filling ports may be squeezed to a greater or lesser degree. To prevent the geotextile tube rolling laterally during filling, the tube has to be (temporarily) secured horizontally (see Figure 5.2). A slight slope in the foundation can lead to tilting or rolling. It should be checked that the horizontal support of the tube does not hinder its sinking when working at some water depth.

The fill material can be pumped into the geotextile tube as a slurry (sand-water mixture) that generally has a ratio of 1 (solid): 4 (water) to 1:5 (based on volume). The surplus water flows through the geotextile skin and via the filling ports out of the tube. It can be determined that if the geotextile tube is filled to around 70% to 80% of its theoretical circular area, the filled height is about half of the 'flat' width of the geotextile tube (width of the part that comes into contact with the bed). A higher filling percentage is also possible although this increases the tensile stresses in the geotextile skin and the tubes are more likely to roll when placed on (slightly) sloping subsoil [22].

Figure 5.2 Temporarily securing geotextile tube.

Figure 5.3 Laying out and pumping up a geotextile tube.

The design of a structure using geotextile tubes depends on the filling procedure used. It is during the filling process that the greatest loads act on the geotextile and seams. The pump capacity and filling speed are key contributory parameters.

5.2.2 Pump speed and pump capacity

The sand (the structural filling material for the geotextile tube) is supplied as a sand-water mixture by a pressure pipe from the pump. In this kind of hydraulic transport environment, pump speed and pipe diameter are dependent on a number of variables, and these are explained below.

The minimum required pump speed is governed by two variables – the sand quality and the pump diameter. It is essential to know in advance of the work what the grain-size distribution is of the sand to be used. A particle size distribution curve can determine the average diameter of the sand.

$$D_{mf} = \frac{D_{10} + D_{20} + D_{30} + D_{40} + D_{50} + D_{60} + D_{70} + D_{80} + D_{90}}{9} \tag{5.1}$$

where:

D_{mf} = average grain diameter [m].

Once the average grain diameter is established, the critical pump speed (v_{cr}) for an appropriate pipe diameter ($\varnothing_{pipe}$) can be determined. Führboter among others [49] developed experimental formulae to determine v_{cr} given different values of D_{mf} and $\varnothing_{pipe}$. Table 5.1 below shows the variation in v_{cr} for different pipe diameters $\varnothing_{pipe}$ and average sand grain diameters D_{mf}.

Because the critical pump speed is the speed where sand deposition in the pressure pipe just does not occur; in practice a minimum pump speed = v_{cr} + 0.5 m/s is used. Higher speeds are not advisable since the required capacity (and thus energy consumption) and wear and tear of the pump increases by the square of the pump speed.

Various techniques are used to feed the sand-water mixture into the sand pump:

- The sand may be supplied from barges or ships and can be gathered from the bottom using a suction dredger. Another option is to use a sand pump: In this case a crane unloads the barges and then deposits the sand into the sand pump.
- The sand is extracted using a cutter-suction dredger or suction dredger and pumped directly into the sand pump. This method generally applies only to small-scale work using a suction dredger of limited capacity. Normally, the capacity of modern dredgers is too high to fill the tubes directly.

Table 5.1 Critical pump speed v_{cr} as a function of sand quality and pipe diameter.

D_{mf} sand in μm	Sand type	$\varnothing_{pipe}$ in m				
		0.30	0.35	0.40	0.45	0.50
100	Fine	2.40 m/s	2.59 m/s	2.77 m/s	2.94 m/s	3.10 m/s
200	Moderately fine	3.09 m/s	3.34 m/s	3.57 m/s	3.78 m/s	3.99 m/s
300	Moderately coarse	3.43 m/s	3.71 m/s	3.96 m/s	4.20 m/s	4.43 m/s

Table 5.2 Sand production Q [m³/hr] as a function of the sand quality, the critical pump speed (based on v_{cr} + 0.5 m/s from Table 5.1) and a sand/water mixture ratio of 20%.

D_{mf} sand in mm	Sand type	$\varnothing_{pipe}$ in m				
		0.30	0.35	0.40	0.45	0.50
0.100	Fine	148 m³/hr	214 m³/hr	296 m³/hr	394 m³/hr	509 m³/hr
0.200	Moderately fine	183 m³/hr	266 m³/hr	368 m³/hr	490 m³/hr	634 m³/hr
0.300	Moderately coarse	200 m³/hr	291 m³/hr	403 m³/hr	538 m³/hr	697 m³/hr

The pump capacity is dependent on a number of variables including pipe diameter and pump speed (stated previously), and also the length of the pressure pipe and the pump resistance that must be overcome. What is ultimately determined using these variables, is the required capacity of the pump to be used during tube filling.

If one assumes a sand concentration of 20% by volume (the normal sand/water mixture ratio), then the sand quantities can be determined based on the critical pump speed v_{cr}, pipe diameter and average sand grading, using Table 5.2.

The choice of equipment to be used must be matched with the scale of the project and to the time available for construction of the project.

It should be noted that in the calculated sand production referred to above no account is taken of the normal losses in the dredging process, namely:

* changing barges under the barge-unloading suction dredger or sand pump;
* adjustments to the cutter-suction dredger or suction dredger;
* weather conditions (wind, wave, current);
* changing geotextile tubes;
* refilling geotextile tubes.

Experience shows that the efficiency factor, through these losses, may fall to 50% of the work, which can increase the gross installation time to twice the net installation time.

5.2.3   Fill material

Geotextile tubes may be filled with silt or sand. In many cases the sand that is locally available will be used. The properties of the sand (grain-size distribution and mixture density) play a key role in the design process, such as in the choice of the geotextile. If the sand has too many fines, then the geotextile tube will consolidate very slowly.

When a sand-water mixture is pumped into a geotextile tube, the sand will settle and the water will flow out through the pores and the outlet ports that are used to fill the tube with sand. The sedimentation rate determines how long it takes for the sand to settle out of the sand-water mixture. When the velocity at which the sand-water mixture is pumped through the tube is known, an estimate can be made of the distance required between the inlet and outlet ports. If this distance is too short, the sand has not settled from the sand-water mixture and will flow out of the tube. In case this distance is too long, the tube will only be partly filled.

A design formula is presented in [29] (for the filling of a hopper) for the sedimentation rate of sand from a sand-water mixture:

$$v_{sed} = \frac{w_0 \cdot c \cdot (1-c)^4}{1-n-c} \tag{5.2}$$

where:

v_{sed} = the sedimentation rate (the rate at which the sand bed rises) [m/s];
n = the porosity of the settled sand (fill material) [–];
w_0 = fall velocity of a single grain [m/s];
c = the concentration of the sand in the sand-water mixture [m³/m³].

This formula applies to a situation involving no internal erosion of the sand layer during deposition and for the situation where the water loss through the sand pores can be neglected. The formula can be used to estimate the sedimentation rate at the start of filling of the geotextile tube. Once the geotextile tube is largely filled, the flow of water through the geotextile tube becomes significantly larger and erosion will occur and the sedimentation rate will decrease causing the filling speed to fall significantly.

To determine w_0 in formula 5.2, Stokes Law is used:

$$w_0 = X \cdot \frac{\Delta \cdot g \cdot D_{mf}^2}{18 \cdot v} \tag{5.3}$$

where:

X = shape factor ([29] suggests 0.7) [–];
Δ = relative density of the fill material (= $(\rho_s - \rho_w)/\rho_w$)) [–];
ρ_s = density of the fill material [kg/m^3];
ρ_w = density of the water [kg/m^3];
D_{mf} = average grain diameter of the sand (formula 5.1) [m];
v = kinematic viscosity of water (= $40.10^{-6}/(20 + T)$ where T, the temperature, is in °C) [m^2/s].

When the output concentration of the mixture, the density and the fall velocity of the grains are known, the sedimentation rate can be determined. Two solutions are possible for the sedimentation rate:

• The sand-water mixture in the geotextile tube is approximately still above the sand bed. In this case the sand concentration in the sand-water mixture is fairly constant and only the thickness of the sand-water mixture reduces.
• There is turbulence in the sand-water mixture, leaving a single concentration over the entire thickness which reduces as more sand from the mixture settles.

In the first instance, the sedimentation rate remains constant. In the second instance, the sedimentation rate will vary depending on the sand concentration, c. In the second instance it is also likely that the sand-water mixture could lead to erosion, though this is not considered any further here.

For the first instance described above, the sedimentation rate is simple to calculate. For various sand types and concentrations the sedimentation rate is given in Figure 5.4. The sedimentation rate theoretically depends on both the grain diameter and the concentration. In practice, because the concentration of sand in the sand-water mixture is less than 0.4, the influence of the sand grain diameter dominates.

For the second instance described above, a numerical simulation is required since the sand concentration changes as does the sedimentation rate. As observed already in Figure 5.4, the sedimentation rate at different sand concentrations below 0.4 remains relatively constant. Also, for the second instance, an estimate can be made of the sedimentation rate using a combination of formula (5.2) and Figure 5.4

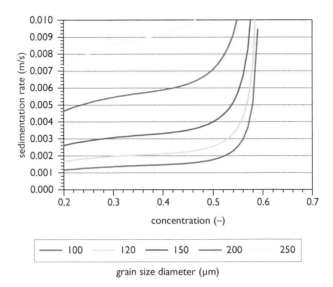

concentration (–)

grain size diameter (μm)

Figure 5.4 Calculation of the sedimentation rate of sand in a geotextile tube.

(which is based on it). This is confirmed by numerical simulations where the initial concentration of the sand-water mixture pumped into the geotextile tube must be entered as the formula concentration.

As long as the geotextile still has a permeability comparable to the permeability of the sand, the sedimentation rate will be higher than according to the calculation, because there is double-sided drainage while the calculation is based on single-sided drainage.

5.3   GEOMETRIC DESIGN

In Figure 2.1 a general design chart is given for geotextile-encapsulated sand elements. The first step in the design process is to establish the functional and technical requirements. This is an area that falls outside the scope of this manual. As already indicated in Section 2.2, it is assumed that the designer is already at the design process stage and has a clear picture of the functional requirements, has a draft design of the entire structure and wants to provide a more detailed design.

The main dimensions of the structure are established first. This is followed by the size of the elements and the construction of the structure based on experimental data, construction feasibility, economic feasibility and application area. Where possible, this is done in consultation with the contractor.

A key component of the design of a structure using geotextile tubes is the choice of the geotextile, which depends on the desired filter properties and the required tensile strength. The filter function requires the geotextile to be adequately sand-tight during use and adequately permeable during installation. In other words, the geotextile has to prevent the sand washing away and enable the water to flow out.

The strength of the geotextile is mainly governed by the loads to which the geotextile tube is subjected during filling. The geotextile must meet the following overall requirements:

- Be sufficiently permeable;
- Be sufficiently sand-tight;
- Be resistant to pressures in the geotextile tube (during filling);
- Be resistant to localised loads (tearing, vandalism);
- Be resistant to UV radiation.

In addition to the choice of geotextile, the stability of the whole structure must be evaluated. Experience has shown that the construction of a geotextile tube structure with several layers can be difficult. It is preferable to design a single-layer structure utilizing geotextile tubes with a relatively large diameter. A multilayer structure is possible but it has to be more broadly erected, which takes up additional space (see figure 5.5).

To make an assessment of the number of geotextile tubes required to achieve a particular design profile, or the tensile loads acting in the geotextile, information must be acquired on the dimensions of the geotextile tube after filling. There are various methods of calculation available (see Appendices D and E) that give an approximation of the (ultimate) shape of the filled geotextile tube and the tensile loads acting in the geotextile.

At a certain degree of filling a minimum height and maximum width can be determined (see Figure 5.6).

An initial estimation of the dimensions and filled shape can be calculated using the design formulae below. Experiments in the Delta flume of Deltares [38] show that the best estimation of the initial shape (without compaction) are found using the Timoshenko method (see Appendix E and [44]). Computer programs such as GeoCoPS [21] also use this method.

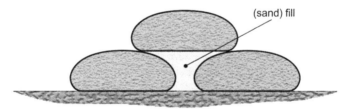

Figure 5.5 Principle of multilayer stacking.

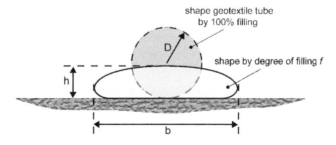

Figure 5.6 Initial estimation of filled geotextile tube dimensions.

$$h \geq (1 - \sqrt{1-f}) \cdot D \tag{5.4}$$

$$b \leq h + \frac{1}{2} \cdot \pi \cdot (D - h) \tag{5.5}$$

where:
h = filled height of the geotextile tube [m];
f = degree of filling with respect to the cross-sectional area (percentage of the area of the theoretical circle) [–];
D = diameter of the geotextile tube at 100% filling [m];
b = filled width of the geotextile tube [m].

Table 5.3 shows the dimensions (and thus the resultant shape) of a geotextile tube for filling levels between 60% and 100% as a function of the radius (R) of the circular diameter of a 100% filled geotextile tube, using the formulae given in Appendix D. The shaded part of the table indicates the most common degree of filling values. Note that these are indicative values and are dependent on the size of the geotextile tube.

Sometimes the term "degree of filling" may be confusing. In this publication (and in the literature), the degree of filling is expressed as a percentage of the theoretical cross-sectional area of a 100% filled geotextile tube. In practice, however, the degree of filling is usually related to the theoretical height of the tube at 100% filling. This is because it is almost impossible to measure the cross-sectional area while filling. As these two values are different, it is therefore advisable to clearly indicate on which parameter the fill rate is based. This is even more important for geotextile containers (to be discussed in Chapter 6), where the degree of filling is sometimes the volume of the geotextile container divided by the volume of the barge that is used for dumping.

Table 5.3 Dimensions and shape of geotextile tube for various levels of filling (tube above water).

f [–]	r [m]	b [m]	h [m]
1.00	1.00 R	2.00 R	2.00 R
0.95	0.70 R	2.28 R	1.59 R
0.90	0.58 R	2.40 R	1.42 R
0.85	0.50 R	2.49 R	1.29 R
0.80	0.43 R	2.56 R	1.17 R
0.75	0.37 R	2.63 R	1.07 R
0.70	0.32 R	2.69 R	0.98 R
0.65	0.28 R	2.74 R	0.89 R
0.60	0.24 R	2.79 R	0.81 R

f = degree of filling.
r = radius of the quadrants on both sides of the straight base of the geotextile tube.
b = maximum width of the geotextile tube.
h = height of the geotextile tube.
R = radius of the theoretical circle for 100% filling ($R = D/2$).

5.4   FAILURE MECHANISMS AND SAFETY CONSIDERATIONS

In designing a structure with geotextile tubes the following failure mechanisms have to be taken into account:

- Rupture of the geotextile tube through excessive pressure during filling;
- Rupture of the geotextile tube through inadequate strength of the seams;
- Instability (rolling) of the geotextile tube during filling;
- Instability of the geotextile tube when subjected to wave attack;
- Instability of the geotextile tube when subjected to current flowing over the structure.

The design formulae for these failure mechanisms are discussed below.

In the design process a safety factor of 1.1 to 1.2 is normally used for geotextile tubes. In addition, strength-reduction material factors are used. The strength-reduction material factors to be applied can best be derived from the material data provided by the manufacturer and the tables contained in this publication.

During the filling of the geotextile tube, the geotextile may be weakened by the abrasion of the sand-water mixture. In practice, a strength-reduction material factor of 1.25 is used. This chapter uses the following (commonly used) values for the strength-reduction factors:

γ for strength of the seams = 2
γ for creep = 1.4
γ for abrasion damage related to the above = 1.25

This leads to an "overall" strength-reduction material factor of 3.5.

5.5   DESIGN ASPECTS

Once the main dimensions and the construction of the structure as well as the size of the elements have been determined, the detailed design is carried out. The structure is assessed in respect of the components shown in the design chart (see Figure 2.1), including the required tensile strength of the geotextile and the stability requirements for waves and water currents. If the required tensile strength is too high for the geotextile and seams to be supplied at an acceptable cost, a smaller geotextile tube or different operating method may be selected, provided the overall design requirements can be fulfilled. Following this, the geotextile tube is checked for stability requirements against waves and water currents, and geotechnical stability associated with any tube stacking. If the stability requirements are not fulfilled either a larger geotextile tube must be selected or supplementary measures taken. The next section looks more closely at the various components of the design cycle.

5.5.1   Material choice and fabrication

A standard geotextile tube is made of woven geotextiles composed of polypropylene or polyester. Woven geotextiles readily retain sand and exhibit a relatively high tensile strength. With low maximum breaking strains (<20%) relatively little deformation occurs upon loading. This is an advantage since the shape of the tube is maintained after filling. Relatively low maximum breaking strains may also mean that tubes may be damaged when subjected to larger deformations, e.g. due to local scour or instabilities. Small diameter tubes are sometimes made of nonwoven geotextiles that allow greater deformation, however, since the tensile strength of nonwovens is lower than that of woven material, large diameter tubes cannot be made and the shape of the tube cannot be maintained over time when using a nonwoven.

The geotextile rolls have a maximum width of 4 to 6 metre. Strips with the width of the rolls and the length of the desired circumference of the tube and are sewn together along the edges to form the tube. Moreover, at regular distances, filling and draining ports are sewn into the geotextile tube structure (see Figure 5.7). Manufacturers have their own proprietary techniques for the precise shape of these ports that govern the maximum pressure that can be applied during the filling of the geotextile tube. In most cases this is approximately 5 kPa. The opening of the drainage port is then 0.5 m or less (depending on the sand concentration in the fill) above the top of the geotextile tube. In the calculations in this chapter 5 kPa is used as the pressure on the upper side of the tube. This pressure determines the shape of the tube and also the maximum degree of filling.

The seams are often the weakest point of the geotextile tube. In sewing the geotextiles together a butterfly seam or (double) J-seam is normally used (see 2.4.5). Proprietary seaming techniques may also be used in order to achieve higher seam efficiencies. The strength of the seams made in the factory is approximately 50% to 80% of the maximum strength of the geotextile materials.

5.5.2 Required tensile strength

The geotextile has to be sufficiently strong to resist the loads that occur during the filling of the geotextile tube. To determine the ultimate load on the geotextile and the

Figure 5.7 Drainage port in action during filling a geotextile tube for Delta flume tests [38].

seams, the degree of filling and resulting shape of the geotextile tube after filling has to be established. The degree of filling depends on the following factors, amongst others:

* whether filling below or above water;
* filling pressure.

The relationship between the shape of a geotextile tube and the tensile load in the geotextile (as a result of the pressure exerted by the fill material) may be derived using the "Timoshenko method", the background of which can be found in Appendix E and [44]. This method may be presented as analytical solutions, but given their complexity can only be solved using numerical methods, e.g. the computer programs GeoCoPS [21] and "Simulation of Fluid Filled Tubes for Windows" (SOFFTWIN) by Palmerton [13].

The analysis method is based on the assumption that there is no transfer of shearing stresses by the fill material on the surface of the geotextile in that part of the tube where there is no contact with the foundation and that the geotextile tube has no flexural stiffness. Under these conditions the tensile load in this part of the geotextile is constant.

The relationship between internal pressure, tensile load and curvature of the geotextile is:

$$T = p \cdot r \tag{5.6}$$

where:
T = tensile load in the geotextile [kN/m];
p = pressure in the fill material [kN/m^2];
r = radius of curvature at a random point on the geotextile skin [m].

For the design calculation of a geotextile tube structure several numerical calculations are made (based on the Timoshenko method) to determine the tensile load in the geotextile, the results of which are shown in Figures 5.8 to Figure 5.10. If the geotextile tube is to be laid entirely under water the following steps must be followed:

* Choose a desired height of the geotextile tube (in this example h = 2.0 m);
* Use Figure 5.9 to determine the corresponding circumference of the geotextile (in this example S = 10.4 m), the degree of filling with respect to the area is 82% and the height ratio compared to the theoretical diameter is 62%;
* Use Figure 5.10 to determine the corresponding theoretical tensile load in the geotextile skin (in this example T = 16 kN/m).

For a geotextile tube of 2 m height above water a corresponding theoretical tensile load of 26.5 kN/m is found. In this case the degree of filling is 72% and the height ratio with respect to the theoretical diameter is 52%.

A safety factor has to be applied to the calculated tensile strength to account for the influence of creep, ageing and the presence of seams. The theoretical required tensile strength, as shown in Figure 5.10, is therefore increased in the design by a factor of 3.5 (see Section 5.4).

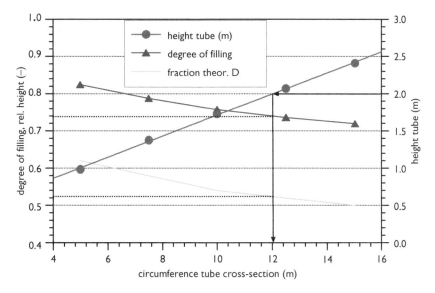

Figure 5.8 Calculated dimensions of the geotextile tube above water as a function of the degree of filling. Assumed unit weight 20 kN/m³ for pressure on the upper side of the geotextile tube during the filling 5 kPa (Note: this figure applies only for 5 kPa. The degree of filling, height and circumference of the geotextile tube are thus related to each other as above).

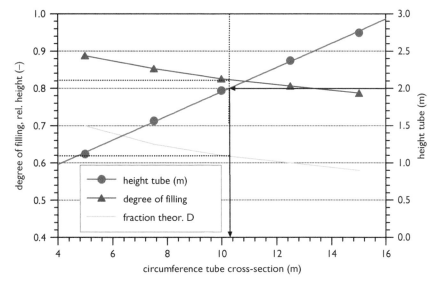

Figure 5.9 Calculated dimensions of the geotextile tube below water as a function of the degree of filling. Assumed unit weight 10 kN/m³ for overpressure on the upper side of the geotextile tube during the filling 5 kPa (Note: this figure applies only for 5 kPa. The degree of filling, height and circumference of the geotextile tube are thus related to each other as above).

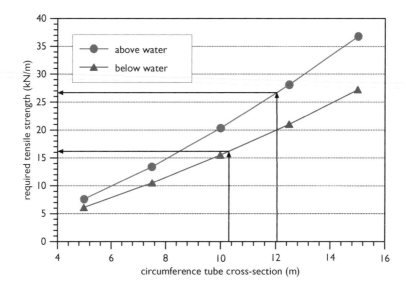

Figure 5.10 Theoretical required tensile strength of geotextile as a function of the circumference of the geotextile tube at a pressure during filling of 5 kPa on the upper side of the tube.

The method described below can be used to provide an approximate estimate of the tensile loads that will occur during the filling process.

Formula 5.6 has to be solved numerically because the shape of the geotextile tube is unknown before filling. This shape can be approximated using Table 5.3 along with the radius of curvature. Then formula 5.6 can be used to determine the tensile load in the geotextile tube.

For example, the radius of the notional circular diameter (R) can be calculated from the value of the tube circumference (S). $S = 10.4$ m $\Rightarrow R = 1.66$ m. The desired height is 2.0 m. This corresponds here with $1.20 \cdot R$. The radius of the quadrants on both sides of the straight basis (r) is, after interpolation, $0.45 \cdot R$ $(= 0.74$ m$)$, with the degree of filling ≈ 0.81.

If the above example is calculated for placement under water, with an overpressure during filling of 5 kPa and a unit weight (under water) of 10 kN/m^3, the pressure in the fill material at the bottom of the geotextile tube is $p_{under} = 5 + 2 \cdot 10 = 25$ kN/m^2. The tensile load is now: $T \approx p \cdot r_1 = 25 \cdot 0.74 = 18.5$ kN/m. The difference with the graphical method is some 10%. This method is therefore suitable to quickly obtain an indication of the required tensile strength for a selected height.

5.5.3   Stability

The shape (and thus also the deformation) of a geotextile tube plays a key role in stability control. It is conceivable that a flat geotextile tube is less likely to roll or tilt than a round one. Also the load on a geotextile in a flat geotextile tube will be lower. Consequently, it may be considered that a low degree of filling is desirable. However, a lower degree of filling means a lower height of the tube and consequently, more tubes will be required to arrive at the design height of the structure, with increased cost. Moreover, it is more difficult to stack geotextile tubes given the likelihood of them rolling or shearing. Therefore an optimum degree of filling must be sought.

During the construction phase, as long as the fill material is still liquid, the geotextile tube is susceptible to tilting. Under these conditions the tube has very little torsional resistance and should be fixed horizontally, but not to the extent that this hinders vertical movement during filling. After construction the geotextile tube has acquired its final shape and this provides the required resistance to wave and current loads.

Research [38] has shown that in this case the degree of filling (i.e. measure of the 'roundness') and the fill material are also important parameters. The design formulae for the stability of geotextile tubes under the influence of waves and currents are given below.

Stability in waves

Based on the available results of research into the stability of geotextile tubes (under wave loading) it can be concluded that the critical wave height (height of the wave where the geotextile tube begins to move) is about the same as the theoretical diameter (the diameter at 100% degree of filling) of the geotextile tube. This only applies to those geotextile tubes most exposed, i.e. those lying around the still water line. This critical wave height is also found from the theoretical equilibrium of forces.

In [22] several stability requirements are stated for geotextile tubes on the crest of a breakwater (around the still water line or submerged) on the basis of small/scale model research [32] carried out on concrete-filled elements. For elements filled with sand, formula 5.7 can be used for an initial approximation.

$$\frac{H_s}{\Delta_t \cdot D_k} \leq 1.0 \tag{5.7}$$

with:
$D_k = l$ (if the geotextile tubes are parallel to the direction of the wave attack and its length is less than 2 times its width$(l < 2 \cdot D_k)$ in the calculation) [m];
$D_k = b$ (if the geotextile tubes are perpendicular to the direction of the wave attack) [m].

where:
H_s = significant wave height [m];
Δ_t = relative density of the geotextile tube [–];
D_k = effective thickness of the geotextile tube [m];
b = width of the geotextile tube [m];
l = length of the geotextile tube [m].

The formula presented above is a simplification. In [41] it is shown that, depending on the circumstances, the stability relations can be more complicated. In the case of a single tube sitting on a relatively smooth, flat and incompressible surface (e.g. a foundation of dense sand) the stability against horizontal sliding decreases significantly. In this case, additional measures should be undertaken to prevent horizontal movement, e.g. placing a threshold or dredging a key trench [41].

The variable b may be approximated as $b = (1.1 \text{ to } 1.2) \cdot D$.

where:
  D = theoretical diameter of a tube [m].

When the crest layer is composed of two tubes connected structurally the equivalent width is equal to $2 \cdot b$, Without structural connections two tubes have no greater stability than one tube [41].

Despite geotextile tubes being able to have a relatively large length (100 m is feasible) it is recommended in formula 5.7 that the length used in the calculations (in the case of the tubes installed parallel to the direction of wave attack) does not exceed twice the average thickness or height (D_k) of the (upper placed) geotextile tube [22].

Stability in water current over the top of the structure

Little information is available on the stability of geotextile tubes under the influence of currents. The equilibrium of forces on a single element suggests the following theoretical stability requirement [24] (see also 3.5.5):

$$\frac{u_{cr}}{\sqrt{g \cdot \Delta_t \cdot D_k}} \leq 1.2 \tag{5.8}$$

where:
  u_{cr} = critical water flow velocity [m/s];
  g = acceleration due to gravity ($g = 9.81$) [m/s^2].

In [9, in Dutch] the results of experiments with 'sand sausages' under various hydraulic loads are shown, with the conclusion that geotextile tubes are more stable than given by the formulae in [22], which found:

$$\frac{\mu_{cr}}{\sqrt{g \cdot \Delta_t \cdot D_k}} \leq 0.5 \text{ to } 1.0 \tag{5.9}$$

This last value (0.5 to 1.0) is recommended as a preliminary design value. If the stability of the geotextile tube is critical with respect to current, it is advisable to do (large scale) model tests.

5.6   CONSTRUCTION ASPECTS

From the previous sections the points listed below are relevant to construction:

• As a result of the permeability of the geotextile, the tube cannot be 'pumped up' with water alone. First some sand will have to be pumped in to restrict the pores. Following this, some pressure build up is possible with the geotextile tube taking up a rounded shape;

- To get an evenly distributed filling, adjacent ports can be opened and closed;
- To prevent the geotextile tube rolling away during filling, the tube must (temporarily) be fixed horizontally. A slight slope in the foundation can cause tilting or rolling. It must be ensured that the horizontal fixing forms no hindrance to the change in geotextile tube shape during filling;
- The machinery used during construction must be chosen with a view of the scale of construction. For example, for a small-scale installation with geotextile tubes a smaller installation set will be required;
- The ratio between the filling time and the total construction time (the construction efficiency factor) can fall to 50% when taking into account changing the barge, changing the suction dredger/ground compressor, adjusting the cutter suction dredger or suction dredger, fluctuating weather conditions (wind, wave, current), changing the geotextile tube, etc. As a result, the construction time is at least twice the filling time;
- A multilayer tube structure is feasible but must be more broadly set up given the added space it requires. The gaps between tubes have to be filled up with sand/fine granular material and it must be ensured that this supplementary fill material does not wash away when the tube lying above is being filled;
- The elements used in a multilayer tube structure must be the same size (insofar as possible) in view of the standardisation of the construction process.

5.7   CALCULATION EXAMPLE

Given:

The example is a dam composed of stacked geotextile tubes (figure 5.11) in a lake. The task is to test the stability of the structure.

Parameters

- Degree of filling of the geotextile tubes: $f = 0.7$ (70%);
- Circumference of the geotextile tubes: $S = 12.3$ m;
- Geotextile tube material: woven polypropylene (PP);
- Maximum tensile strength of the geotextile: $T_m = 80$ kN/m;
- Pore size of the geotextile tube: $O_{90} = 250$ µm;

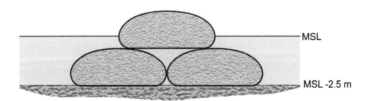

Figure 5.11 Cross-sectional sketch of the dam structure.

- Sand gradation: D_{10} = 100 μm, D_{20} = 140 μm, D_{30} = 178 μm, D_{40} = 216 μm D_{50} = 255 μm, D_{60} = 292 μm, D_{70} = 308 μm, D_{80} = 312 μm en D_{90} = 340 μm;
- Porosity of sand in tube: n = 0.4;
- Sand density: ρ_s = 2650 kg/m³;
- Water density: ρ_w = 1000 kg/m³;
- Acceleration due to gravity: g = 9.81 m/s²;
- Diameter of sand pump: 0.400 m;
- Summer level: MSL – 0.2 m;
- Winter level: MSL – 0.4 m;
- Significant wave height: H_s = 0.8 m;
- Sand concentration in mixture: 20%.

Solution

Pump speed and pump capacity

The first step is to determine the critical pump speed v_{cr}. For this purpose, the D_{mf} has to be established first according to 5.2.2:

$$D_{mf} = \frac{D_{10} + D_{20} + D_{30} + \cdots + D_{90}}{9}$$
$$= \frac{100 + 140 + 178 + 216 + 255 + 292 + 308 + 312 + 340}{9} = 238 \; \mu m$$

v_{cr} can now be determined using Table 5.1. At a D_{mf} = 0.238 mm and a pipe Ø of 400 mm, v_{cr} = 3.72 m/s and the pipeline speed V_{pipe} = v_{cr} + 0.5 = 3.72 + 0.5 = 4.22 m/s is recommended. The corresponding sand production is determined from Table 5.2 ⇒ Q = 381 m³/hour. This is a theoretical value. During the filling process it will be necessary to vary the sand concentration of the sand-water mixture and the related flow rate to fill the tube evenly. The presented value is indicative.

Sedimentation rate

A sand-water mixture with a sand concentration of 20% is assumed along with a water temperature at 12°C. The sedimentation rate can be determined from formulae (5.2) and (5.3):

$$v = \frac{40 \cdot 10^{-6}}{20 + 12} = 1.25 \cdot 10^{-6} \, m^2/s$$

$$\Delta = \frac{2650 - 1000}{1000} = 1.65$$

$$w_0 = X \cdot \frac{\Delta \cdot g \cdot D_{mf}^2}{18 \cdot v} = 0.7 \cdot \frac{1.65 \cdot 9.81 \cdot (238 \cdot 10^{-6})^2}{18 \cdot 1.25 \cdot 10^{-6}} = 2.85 \cdot 10^{-2} \, m/s$$

$$v_{sed} = \frac{w_0 \cdot c \cdot (1 - c)^4}{1 - n - c} = \frac{2.85 \cdot 10^{-2} \cdot 0.2 \cdot (1 - 0.2)^4}{1 - 0.4 - 0.2} = 5.8 \cdot 10^{-3} \, m/s$$

Dimensions and shape

For the lower tube elements the theoretical maximum radius can be determined from the known circumference:

$$R = \frac{S}{2 \cdot \pi} = \frac{12.3}{2 \cdot \pi} = 1.96 \text{ m}$$

Based on a degree of filling of 70%, it follows from Table 5.3 (not based on the more accurate Timoshenko method, see 5.5.2) that:

$r_{70\%} = 0.63$ m
$b_{70\%} = 4.27$ m
$h_{70\%} = 1.92$ m

Filling time and filling length

Based on the sand capacity of 381 m³/hour, the length of tube that will be filled within an hour can be estimated. The diameter of the tube is assumed to be 100% filling $\pi \cdot R^2 = 12.1$ m². For a degree of filling of 70% this is 8.4 m². Over 1 hour the tube will be filled up to 381/8.4 = 45 m. This is a simplified relationship. In actual fact the tube will be filled quicker where the sand is introduced, while further on it will take longer before the tube is full.

Note: In general the total tube cross-section will not be available during filling (8.4 m²), but only a proportion (maybe only ¼) because the cross-section will be partly filled with sand.

Required tensile strength

The geotextile tubes have a circumference of 12.3 m and lie entirely under water. From Figure 5.9 a tensile strength of 20 kN/m can be determined. The initial degree of filling is then around 80%, but it is assumed that this degree of filling is necessary to reach 70% when the filling stops. Due to strength-reduction material factors the required tensile strength will be raised by a factor of 3.5 to arrive at a design strength (see 5.4) ⇒ required design tensile strength = 20 · 3.5 = 70 kN/m. Use is made of a woven polypropylene geotextile with a maximum tensile strength of 80 kN/m. This provides an overall safety factor of 80/70 = 1.14, so the requirement for the tensile strength is fulfilled.

Sand density

Table 2.3 presents the formula for the required pore size (O_{90}) of the geotextile. For dynamic load (wave attack) the requirement is:

$$O_{90} < 1.5 \cdot D_{10} \cdot C_u^{1/2}$$

$$1.5 \cdot 100 \cdot \left(\frac{292}{100}\right)^{1/2} = 256 \ \mu\text{m} \ \Rightarrow \text{complies}$$

Stability in waves

For the stability regarding waves, the element on top is considered in terms of formula (5.7) and Table 5.3.

$$\frac{H_s}{\Delta_t \cdot D_k} \leq 1.0$$

$$\frac{0.8}{0.98 \cdot 1.92} = 0.43 \Rightarrow \text{complies}$$

Stability in current

The main influence on current in a large lake will be caused by the wind, the magnitude of which is unknown. So we look at the maximum allowable current that may be present before instability occurs according to formula (5.8):

$$\frac{u_{cr}}{\sqrt{g \cdot \Delta_t \cdot D_k}} \leq 1.2 \Rightarrow u_{cr} = 1.2 \cdot \sqrt{g \cdot \Delta_t \cdot D_k} = 1.2 \cdot \sqrt{9.81 \cdot 0.98 \cdot 1.92} = 5.1 \text{ m/s}$$

It is highly unlikely that this current will occur in a lake due to wind effects. The geotextile tubes thus comply also in terms of stability due to water currents.

Chapter 6

Geotextile containers

A geotextile container is a large geotextile-encapsulated sand element containing 100 m³ to 800 m³ sand and is dropped through water from a split barge.

6.1 APPLICATION AREAS AND GENERAL EXPERIMENTAL DATA

Geotextile containers can be used in the following applications:

- To protect structures in water against scour holes by filling any scour holes with the containers;
- As core material of a breakwater that is covered with armour rock;
- Raising the bed under the core of a breakwater;
- As offshore sills for beach nourishment projects;
- As offshore artificial reefs;
- As offshore submerged dykes;
- Quay structures made of geotextile containers, or pressure relief walls behind quays;
- Construction of containment dams.

Applications in the Netherlands include bank protection along the Oude Maas in Rotterdam, bed and bank protection for the sailing route at Harlingen, submerged breakwater at the Botjes Zandgat sand extraction site and the dam in the Cornelis Douwes canal. These projects are described in Appendix B of [14].

6.2 INSTALLATION PROCEDURE

6.2.1 General

A geotextile container is placed in the hopper of a split barge and it is then filled with sand (mechanically or hydraulically). The open top of the geotextile container is then folded and sewn tight. The split barge is then positioned at the correct dropping point. Finally, the split barge is opened and the geotextile container is dropped in the water (see Figure 6.1).

Figure 6.1 Application of a geotextile container (filling, closure, releasing, falling).

6.2.2   Fill methodology

A geotextile container can be filled with sand mechanically (by grab bucket) or hydraulically (by pumps). Within the entire construction cycle filling is the major time factor and so has to be performed relatively quickly. Dry filling is preferable to hydraulic filling because it enables even filling and more air is contained in the fill material. The significance of the air is discussed in 6.5.3. Dry filling also has the advantage of enabling a steeper sand profile. This generates less width for the same filling volume, and enables the geotextile container to move more easily through the opening of the split barge during the drop. Filling must be spread evenly over the length of the geotextile container. If not, during the drop, the geotextile container will shift unevenly through the split barge opening, with one end passing through and the other end being impeded, thus increasing the tensile stresses in the geotextile container (see Figure 6.2).

6.2.3   Fill material

The grain-size distribution and density of the fill material play a key role in the design process and the choice of the geotextile skin. Published literature [42] also refers to geotextile containers that have been filled with dredged sediments for regulated storage. However, in this manual the geotextile container is considered purely as a structural element and always filled with sand.

The use of (clean) sand has the advantage that due to internal friction the material can absorb more of the drop energy that is released during container impact on the bottom. This reduces the possibility of geotextile rupture significantly.

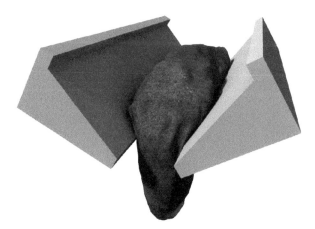

Figure 6.2 Sketch of a partially impeded geotextile container during the drop.

6.2.4 Degree of filling

The ultimate shape of a geotextile container depends mainly on the degree of filling; theoretically 100% filling will result in a circular container. This manual, defines the degree of filling of a geotextile container as the relationship between the actual surface of cross-section of the container (perpendicular to the length) and its largest possible surface of cross-section (a circle). The formula is as follows:

$$f = \frac{4 \cdot \pi \cdot A}{S^2} \qquad (6.1)$$

where:

f = degree of filling [–];
A = filled cross-sectional area of the geotextile container [m²];
S = circumference of the geotextile container [m].

In Figure 6.3 the circumference is shown as a function of the degree of filling and the cross-section. The grey shaded region shows the degree of filling between 20% and 80%. This region is the most common for geotextile containers and tubes. From 20 to 50% are usual values for containers, from 50 to 80% for tubes.

The degree of filling of a geotextile container is defined in the literature in two different ways. The common way (as shown above) is expressed as a function of the circumference and the filled cross-sectional area of the geotextile container, but sometimes it may be defined as a function of the volume of the split barge. It should be noted that these two definitions differ significantly. Thus, a degree of filling of 46% can, according to the first definition, correspond to a degree of filling of 80% based on the second definition. This difference is due to the shape of the split hopper bin in cross-section. When in literature fill percentages of 70% and higher are stated, one can assume that the degree of filling is calculated in relation to the volume of the split barge and not according to formula 6.1.

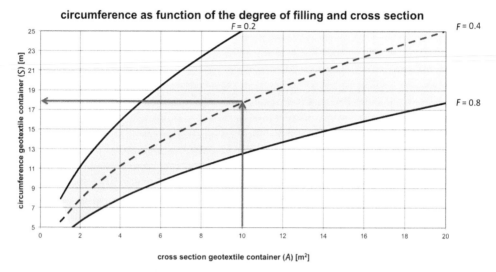

Figure 6.3 Relationship between degree of filling, filled cross-sectional area and circumference of geotextile containers. For example, a geotextile container with a cross-sectional area of 10 m² with a degree of filling of 40% (*f* = 0.4) will have a circumference of about 18 m.

6.3   GEOMETRIC DESIGN

In Figure 2.1 a general design chart is given for geotextile-encapsulated sand elements. The first step in the design process is to establish the functional and technical requirements. This is an area that falls outside the scope of this manual. As already indicated in Section 2.2, it is assumed that the designer is already at the design process stage and has a clear view of the functional requirements, has a draft design of the entire structure and wants to elaborate on it.

First, the main dimensions of the structure are established. Then, the size of the elements and the construction of the structure based on experimental data, construction feasibility, economic feasibility and application area are carried out. Where possible, this is done in consultation with the contractor.

In [14, in Dutch] indicative values are given for the ultimate dimensions of geotextile containers after placement in relation to the size of the split barge. Table 6.1 shows a revised version of this table based on a degree of filling of around 40% of the maximum possible cross-sectional area and of around 80% of the bin volume of the split barge used in placement. The shaded part gives an indication of common split barge dimensions.

The filled cross-sectional area (*A*) of the geotextile container can be calculated by multiplying the width (*b*) by the height (*h*) of the geotextile container (*b* · *h*), but that will overestimate the filled cross-sectional area because the shape is not rectangular. A more accurate way of determining *A* is to divide the volume (*V*) of the container by the length (*l*) of the barge bin (*V/l*).

Table 6.1 Relation between the size of a split barge and the placed dimensions of a geotextile container for a geotextile container with a degree of filling of 40% that fills 80% of the barge bin. The width given is an indication since the width changes with the height. This table is a revised version of the table from [14, in Dutch].

Split barge		Geotextile container dimensions after placement				
Hold contents (80% filled) [m³]	Hold length [m]	Filling [m³]	Circumference [m]	width [m]	A (V/l) [m²]	Height [m] deviation ±25%
200	23.4	160	14.5	6.7	6.8	1.25
250	25.2	200	15.7	7.2	7.9	1.40
300	26.8	240	16.6	7.6	9.0	1.50
350	28.2	280	17.5	8.0	9.9	1.55
400	29.5	320	18.3	8.4	10.8	1.60
450	30.7	360	19.0	8.7	11.7	1.65
500	31.7	400	19.7	9.0	12.6	1.75
550	32.8	440	20.4	9.3	13.4	1.80
600	33.7	480	21.0	9.6	14.2	1.85

In some cases, the height of the dropped geotextile container is important, for example where a dam is built from these elements and where a distinction has to be made between containers dropped in shallow water (less than 8 m depth) and in deep water (more than 8 m depth). The cross-sectional shape of a geotextile container dropped in deep water can be assumed, as a first approximation, to form a rectangle. For the filled cross-sectional area of the bin this is equal to the filling volume (V) divided by the bin length (l). By dividing this filled cross-sectional area by the width (b) of the geotextile container a more realistic height is found when dropping in deep water. The method for calculating the height is generally not very accurate since the height realised depends on many factors (drop procedure, sand grain diameter and sand saturation). So, this must be checked and the procedure adjusted where necessary (for example, if possible a higher degree of filling if the height is too low). Due to these aspects it is difficult to determine the final shape of the geotextile container when dropped in shallow water. The geometry of the installed geotextile container could be determined by using a (multi-beam) echo sounder after installation. It is recommended to determine the shape after every container dropped. In this way it is possible to monitor the shape and position of every individual container.

If a certain cross-sectional area has to be filled with geotextile containers in a hydraulic structure, the values shown in Table 6.1 cannot be used to calculate the filled cross-sectional area of an individual geotextile container. The geotextile containers will take on such a shape during the construction phase that the whole cross-sectional area will be filled. In order to estimate the required number of geotextile containers the filled cross-sectional area per geotextile container has to be calculated. This is determined by dividing the filling volume (V) by the bin length (l). The (minimum) required number of geotextile containers can be determined from the cross-sectional area of the structure divided by the calculated filled cross-sectional area of the geotextile container (see Figure 6.4).

During the drop a geotextile container has to be able to deform in order to slide through the bottom opening of the split barge. For the same volume of fill material the greater the circumference, the more the geotextile container can deform. The requisite circumference of the geotextile container is as follows:

$$S \geq 2.5 \cdot \left(\frac{A}{b_0} + b_0 \right) \tag{6.2}$$

where:

S = circumference of the cross-section of the geotextile container [m];

b_0 = width of the opening of the split barge [m];

A = filled cross-sectional area of the geotextile container in the barge or cross-section of the barge [m²].

The accuracy for determining S lies in the order of ±25%. The deviation is due to the factor of 2.5 in Equation 6.2 not being known precisely. In the case of a split barge with relatively high friction during release, the factor of 2.5 should be replaced with a factor of 3 [22]. This will result in a flatter and lower geotextile container after dropping.

In Figure 6.5 the minimum required circumference of a geotextile container (filled with sand) is given for various opening widths of the split barge (b_0) as a function of the filled cross-sectional area.

For a split barge, suitable for dropping geotextile containers, a rule of thumb is that the bin width is equal to 1/5 of the bin length. In addition, the opening width of the split barge should be at least 50% of the bin width.

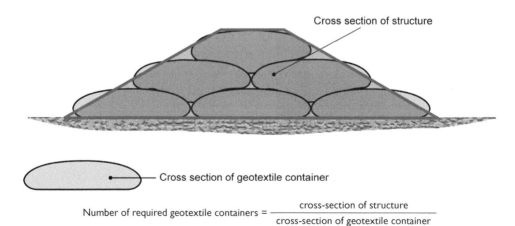

Figure 6.4 Estimation of the required number of geotextile containers in a structure.

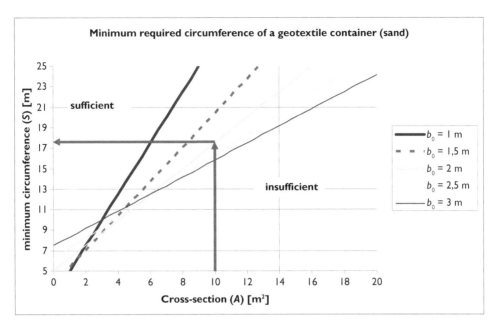

Figure 6.5 Minimum required circumference of a geotextile container (according to formula 6.2) as a function of the filled cross-section and opening width of the split barge. For example, a split barge with a maximum opening width of 2 m, a geotextile container filled with sand and a filled cross-section of 10 m², a minimum circumference of 17.5 m is required.

6.4 FAILURE MECHANISMS AND SAFETY CONSIDERATIONS

For a structure using geotextile containers, the following failure mechanisms have to be considered:

- weakening/rupture of the geotextile and/or seams on the geotextile container leaving the split barge;
- rupture of the geotextile and/or seams upon container impact on the bottom;
- failure of the geotextile under the action of waves and/or currents;
- loss of stacking stability of the geotextile containers;
- damage due to construction works;
- damage due to any surface finishing (for example during installing an armour layer);
- sand migration within or from the geotextile container;
- draining retained material (subsoil/sand behind the geotextile containers) when using geotextile containers as bank protection;
- sand scour from underneath the geotextile containers due to water currents or wave action;
- degradation of the geotextile (creep, UV radiation, etc).

The installation phase governs many of the required mechanical properties of the geotextile container. Normally, during construction a safety factor of 1.1 to 1.2 is applied. Since the aging of the geotextile doesn't have to be taken into account in the design phase and most of the loads are only present during construction, this is generally equivalent to a safety factor of 3 during use.

6.5   DESIGN ASPECTS

Once the main dimensions, the construction of the structure and the size of the elements have been determined, the detailed design is carried out. The structure is assessed in respect of the components shown in the design chart, including the required tensile strength for the geotextile and the stability requirements for waves and currents.

In most cases the energy that the geotextile has to absorb during impact governs the strength of the geotextile. This energy is absorbed by the tensile strength and the maximum strain of the geotextile. If the requirements for tensile strength and/or strain are so high that the geotextile is too expensive, then a lower degree of filling, smaller geotextile containers or another construction method can be chosen. Also, the geotextile containers have to be assessed for stability against waves and currents and for stacking stability. If these stability requirements are not fulfilled, then either a heavier geotextile container must be selected or supplementary measures taken. The next section looks more closely at the various components of the design cycle.

In addition to the points above the availability and applicability of split barges must be considered. For example, it may be difficult to obtain split barges of the required dimensions and opening widths due to market demands.

6.5.1   Material choice

Woven geotextiles used for geotextile containers have the same strength in two directions. This strength can vary from 80 kN/m to 200 kN/m. Nonwoven geotextiles may be also used as geotextile containers, but for lower filled container volumes. Here, the tensile strength varies between 30 and 50 kN/m. This is much less than the strength of a woven geotextile, but the much larger elongation at break enables it to absorb a considerable amount of energy during impact. The low strength and larger elongation results in a more deformable container shape than a geotextile of woven material and consequently a flatter container shape after installation. The choice between the two types of geotextiles depends on cost, loads expected in the geotextile during opening and dropping and local experience. Here, the focus is on the use of woven geotextiles because this is the experience in The Netherlands.

The production of the woven geotextile (polyester and/or polypropylene) occurs in factory-controlled quality conditions. In Europe the geotextiles used are CE certified.

6.5.2   Loading on geotextile while opening
the split barge

During opening of the split barge, the geotextile container is pulled down by the weight of the fill material, thus creating frictional forces along the walls of the split barge that result in a tensile load in the geotextile. At the start of the opening these tensile loads are only present in the lowermost part of the geotextile, but just before the geotextile container leaves the barge it covers a large part of the geotextile surface, with only the uppermost part being virtually free of loading.

De Groot [36] has evaluated the tensile load in a geotextile during the opening of the split barge. In most cases, however, this loading is not critical and therefore a simple formula was developed for a first approximation [4, 15]. The maximum possible tensile load in the geotextile during the opening of the split barge is:

$$T = 0.45 \cdot \left(\rho - \rho_w\right) \cdot g \cdot A \tag{6.3}$$

where:
T = tensile load in the geotextile [N/m];
ρ = density of the geotextile container [kg/m³];
g = acceleration due to gravity [m/s²];
ρ_w = density of water [kg/m³];
A = filled cross-sectional area of the geotextile container [m²].

The basis of this formula is the assumption that if just 10% (or less) of the geotextile container lies in the split barge, the container will certainly fall through the opening, at which point the geotextile is subject to the calculated load being 0.9/2 = 0.45 times the submerged weight of the container. If, according to formula (6.3), the tensile load is less than the tensile strength of the geotextile, then the load on the geotextile during the opening of the split barge no longer needs further consideration. Application of a safety factor of 1.1 to 1.2 to this design value is recommended.

Given that the longitudinal seams in a geotextile container should be on the upper side and that this part of the container is the last to leave the barge during the drop, it is not necessary to account for the lower seam strengths in the above formula.

If the loads generated during opening of the split barge approximate the strength of the geotextile, and compliance with formula (6.3) is not a straightforward matter, a more advanced determination of the various loads on the geotextile can be undertaken. In Appendix F determination of the tensile load in the geotextile at each installation stage is described.

Based on experience, the following rule for the opening of the barge can be used: during the drop of the geotextile containers, the split barge must have obtained an opening of 50% of the total width of the barge bin within 20 to 30 seconds. In practice, this requirement is not always achievable and even then the barge can still be used although there is more chance of an uneven drop of the container.

6.5.3 Loading on geotextile upon impact on the bottom

When the geotextile container impacts the bottom, the geotextile is subject to a significant increase in load and deformation. This has to be determined as accurately as possible since this is often the critical criterion with regard to the required tensile strength of the geotextile.

The container fall energy is normally dissipated in two ways: by the strength and strain of the geotextile and by the deformation of the sand in the container. If the foundation of the bottom consists of soft clay, a significant part of the fall energy will also be dissipated by the deformation of the bottom foundation. Such a situation is not accounted for here.

No failure occurs if:

$$E_{fall} < E_{fill} + E_{geo} \tag{6.4}$$

where:

E_{fall} = fall energy per unit of length [J/m];
E_{fill} = maximum energy to be dissipated by the fill material per unit of length [J/m];
E_{geo} = maximum energy to be dissipated by the geotextile per unit of length [J/m].

All the components are expressed here in terms of energy per unit of length, which presents no problem if the geotextile container actually impacts flat on the bottom. However, if a container hits the bottom with its front or back end first, as is shown in figure 6.6, a total energy approach may be warranted.

In the case of a non-parallel drop, most of the fall energy of the entire container has to be dissipated by the geotextile and the sand in only a part of the container. This can increase the energy to be dissipated on a per unit of length basis significantly. Therefore, the greatest possible care should be taken to ensure the container falls as flat as possible on the bottom.

Energy dissipation by the geotextile

The amount of energy that can be dissipated by the geotextile is a function of the maximum strain and tensile strength of the geotextile [22] [17].

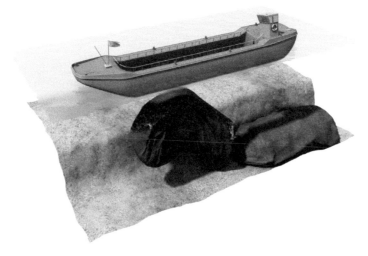

Figure 6.6 Simulated fall of geotextile container from split barge. Note that during the fall the container height is greater than the width. This depiction is only of qualitative value since the behaviour of the geotextile container here is simplified.

$$E_{geo} = \frac{1}{2} \cdot S \cdot J \cdot \varepsilon_m^2 \qquad (6.5)$$

where:

S = circumference of the geotextile [m];
E_{geo} = maximum energy to be dissipated by the geotextile per unit of length [kJ/m];
ε_m = maximum strain of the geotextile [–];
J = tensile stiffness of the geotextile [kN/m].

Formula (6.5) can also be rewritten as:

$$E_{geo} = \frac{1}{2} \cdot S \cdot T_m \cdot \varepsilon_m \qquad (6.6)$$

where:

T_m = tensile load in the geotextile [kN/m];
ε_m = tensile strain in the geotextile [–].

This formula assumes a linear load-strain behaviour and since, in practice, the geotextile reacts less stiffly initially and more stiffly at greater strains, this is an unsafe approximation, but acceptable since the assumption that all kinetic energy is dissipated by the geotextile is on the safe side. At maximum tensile strength the stiffness decreases again and the formula may become conservative. However, that value is never reached in practice since the seams will break before this value is reached.

The strength of the seams and the fastening quality of the closing on top of the container are normally the limiting factors. If the tensile strength of the seams is half that of the actual geotextile (see Table 2.5), the ε_m is normally also half of the value for the entire geotextile (and the energy that can be dissipated is thus just a quarter of the geotextile itself). This demonstrates that the quality of the seams for geotextile containers is crucial. For calculation, it is recommended to use a safety factor of 1.1 to 1.2 on the design value. It should also be noted from this formula that the circumference (S) is directly related to the amount of energy that can be dissipated. Thus, a larger geotextile circumference will dissipate a larger amount of energy.

Energy dissipation by the sand in the container

Calculations in [5] reveal that the fall energy is too great to be fully dissipated by strain in the geotextile alone. Therefore, the component E_{fill} has to contribute a significant level of energy dissipation. The magnitude of this component is a function of the dimensions of the geotextile container and the properties of the fill material. According to [17] the following approximate relationship applies for frictional fill material:

$$E_{fill} = 2 \cdot \sigma_i' \cdot \tan \phi_s \cdot \sqrt{A} \cdot (\sqrt{A} - h) \qquad (6.7)$$

where:

E_{fill} = energy dissipated by the fill material per unit of length [kJ/m];
h = height of the container when dumped [m];

σ_i' = average effective stress of the sand in the container [kPa];
ϕ_s = angle of internal friction of the sand in the container [degrees];
A = filled cross-sectional area of the geotextile container [m²].

The magnitude of this component is only significant for a high value of the effective stress σ_i' and the resulting friction that occurs between the sand grains. Two circumstances can be distinguished:

- a situation where the sand is close to full saturation;
- a situation where there is still plenty of air in the voids.

These two situations are discussed below. However, it will become clear that the energy dissipation in the contained sand depends on other factors (as well) that cannot always be calculated, although guidelines can be given to optimize the energy dissipation in the sand layer.

Influence of the level of saturation

If the condition occurs where the pores are saturated with water an effective stress may arise during impact. This arises partly due to the effect of dilatancy and partly due to the impact itself. During dilatancy the grain skeleton increases in volume, under a shearing deformation, with a resulting tendency to absorb water. Because of the speed of the deformation and the low water permeability of the fill that hinders the flow of water into the dilating voids, negative pore pressures occur in the contained fill voids. A precondition for dilatancy is a sufficient high density of the sand. This precondition is normally not fulfilled, because the sand has been deformed to a large extent when the container leaves the split barge and therefore the density will be close to the critical density. However, some dilatancy has been observed through modelling and field measurements, though it is by no means always present. During impact, there is an increase in total stress. From measurements it appears that this is not taken by the pore pressure, but mainly by the grain skeleton increasing the effective stress directly. In field measurements at the Kandia dam [7] two instrumented containers were dropped and the effective stresses measured. For one with saturated sand the effective stresses remained limited, because the impact was mainly taken by the pore pressures. The other, which was made with unsaturated sand originally, showed an increase in effective stress during impact of 30–100 kPa. 25 kPa of this increase in effective stress was caused by dilatancy, the rest was caused by direct increase of the total pressure during the impact. The potential for dilatancy to occur is greatest if the sand has few fines: the percentage of fines smaller than 63 μm should be less than 20%.

Influence of enclosed air

The influence of air enclosed in the voids of the fill material (unsaturated sand) has been studied in [18]. The conclusion is that the enclosed air increases the undrained shear strength. During the fall the external pressure increases, and since the air in the fill material is compressible the pore pressure will hardly increase, and thus the increase in external pressure will cause the effective stresses to increase. The fill material thus temporarily contributes a greater shear resistance against the deformation caused by impact, and thus reduces the load generated in the geotextile. This has also been termed the 'Bezuijen effect' [15]. Since very little water can flow into the voids during the very short time that the container falls, the increasing external water pressure will result in an increase in the effective stress that is approximately equal to the water pressure at depth, e.g. around 100 kPa at 10 m water depth. The increase in the effective stress due to the impact has to be added to that.

The field measurements at the Kandia Dam did not reveal the 'Bezuijen effect' but showed for the container loaded with unsaturated sand a lot of the impact is 'taken' by the effective stresses. This causes energy dissipation in the fill which was not the case for the container with saturated sand.

To generate this effect requires the use of unsaturated sand-fill. When sand is introduced into the bin the level of saturation will increase but not to full saturation throughout the fill. To limit the load on the geotextile upon impact the use of unsaturated sand also has the advantage that the submerged weight of the geotextile container is lower, making the fall velocity and fall energy less than if saturated sand were used. During filling the split barge the lowermost part of the sand will reach a saturation of 80 to 90%. Some air will always be entrapped in the grain voids. Field measurements have also revealed that air retained in the fill material reduces the loading during impact. (see Figure 6.7 and Figure 6.8). It is important that the air that escapes from the fill material can escape from the geotextile container. During one of the field measurement trials air remained between the fill material and the geotextile, and when the container was dropped, the air moved upwards causing the geotextile container to fall at an angle.

Figure 6.7 Field measurements Kandia dam: Filling of the first instrumented geotextile container for which no effective stress was measured during the fall since the fill material is saturated with water.

Figure 6.8 Field measurements Kandia dam: Filling of the second instrumented geotextile. The unsaturated filling resulted in limited load being exerted on the geotextile during the fall.

Fall velocity and fall energy

The fall energy E_{fall} that has to be dissipated is kinetic energy. This can be written as:

$$E_{fall} = \frac{1}{2} \cdot \rho \cdot A \cdot v^2 \tag{6.8}$$

where:

E_{fall} = fall energy per unit of length [J/m];
v = fall velocity of the geotextile container just before impact [m/s];
A = filled cross-sectional area of the geotextile container [m²];
ρ = density of the geotextile container [kg/m³].

The key parameter in determining the load upon impact on the bottom is the fall velocity of the geotextile container. In Appendix G the development of the fall velocity of the geotextile container during free fall is computed. The maximum fall velocity, the so-called terminal velocity (v_∞), is [22, paragraph 6.3]:

$$v_\infty = \sqrt{\frac{2 \cdot V}{A_s \cdot C_d} \cdot \frac{\rho - \rho_w}{\rho_w} \cdot g} \tag{6.9}$$

where:

v_∞ = terminal velocity [m/s];
V = volume of the geotextile-encapsulated sand element [m³];
A_s = cross-sectional area in the horizontal plane [m²];
C_d = drag force-coefficient (approximately 1) [–].

The terminal velocity is theoretically only reached at a certain depth. In Figure 6.9 the fall velocity is calculated, using the formulae in Appendix G, as a function of

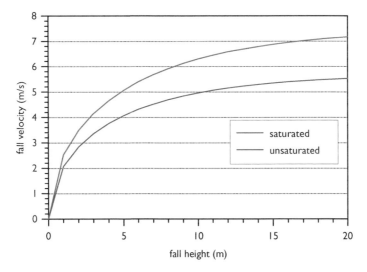

Figure 6.9 Fall velocity as a function of the fall height (=the water depth minus the draught of the split barge) for a container with a volume of 260 m³, saturated fill ($\rho = 1800$ kg/m³) and unsaturated fill ($\rho = 1460$ kg/m³).

the water depth for a container with $V = 260$ m³ and for the cases of saturated fill ($\rho = 1800$ kg/m³) and unsaturated fill ($\rho = 1460$ kg/m³), for $C_d = 1$ and $A_s = 72$ m². For this example, both cases show that terminal velocity still has not been reached at 20 m water depth.

6.5.4 Required tensile strength

Using the analysis presented above, the required tensile strength of the geotextile container can be determined.

If it is assumed that E_{fill} is fully mobilised even at small strain in the geotextile, then the energy absorbed by the geotextile (E_g) is equal to the fall energy minus the energy absorbed by the fill material:

$$E_{geo} = E_{fall} - E_{fill} \tag{6.10}$$

From E_{geo} the strain in the geotextile (ε_g) during impact can be calculated as [22, 17]:

$$\varepsilon_g = \sqrt{2 \cdot \frac{E_{geo}}{S \cdot J}} \tag{6.11}$$

Thus, producing a geotextile tensile load of:

$$T_g = J' \cdot \varepsilon_g \tag{6.12}$$

This tensile load has to be less than or equal to the required tensile strength of the seams in the geotextile container.

When the required impact tensile load is greater than the tensile load calculated for release from the split barge, then the impact condition governs the required tensile strength of the geotextile container. The strength of the seams is the weakest component and this will govern the tensile strength and the corresponding strain. This has a major influence on the result of formula (6.6) (see also the calculation example at the end of this chapter).

6.5.5   Stability in waves

In [5] the development of design formulae for the stability of geotextile containers under wave attack have been given. In a small-scale (1:20) model two different structures were tested, both of which were stacked at a slope of approximately 1(V):3(H), with the difference being the degree of filling of the geotextile containers (46% versus 70%). Based on the results of this model study the following design formula can be given for the stability under wave load of individual geotextile containers with $b/D_k > 4$:

$$\frac{H_s}{\Delta_t \cdot D_k} \leq 2 \quad \text{if} \quad \frac{b}{D_k} > 4 \tag{6.13}$$

where:
   H_s = significant wave height [m];
   Δ_t = relative density of the geotextile container [–];
   D_k = effective thickness of the geotextile container after drop [m];
   b = width of the geotextile container after drop [m].

In [22, see paragraph 5.4.11] a comparable ratio is given for the stability of geotextile containers under the influence of wave attack:

$$\frac{H_s}{\Delta_t \cdot D_k} \leq F \tag{6.14}$$

where:
   $F = 1$ for geotextile containers on the crest of the structure and lying less than the depth of H_s under water, as in tidal areas [–];
   $F = 2$ for geotextile containers that lie lower than the depth of H_s beneath the still water line [–].

During large scale physical model tests on the stability of geotextile containers [37], performed by Deltares, it was observed that besides sliding, the 'caterpillar' mechanism (caused by the migration of sand in the container) is a significant mechanism. The stability of the geotextile containers in the large scale tests was lower than the stability determined in other (smaller) scale model tests. This is explained by the migration of the sand being better modelled in large scale tests than in small scale tests.

Van Steeg and Klein Breteler [37] have concluded the following; the application of geotextile containers is limited to areas with low wave attack ($H_s < 0.75$ m with the crest on the waterline and $H_s < 1.1$ m with the crest $0.75 \cdot H_s$ under water) or in areas with larger waves if the crest is $2H_s$ or more under water. If these criteria are not fulfilled it is possible that there will be sand migration in the container thus leading to undue deformations. It should be noted that this conclusion depends on the degree of filling. The authors have used a low degree of filling (~44%) for the basis of their conclusions. With a higher degree of filling a container should be able to withstand higher wave attack, but usually it is difficult to dump a geotextile container with a higher degree of filling because of the opening width of the barge. The wave attack criterion presented by van Steeg and Klein Breteler is rarely a limiting factor since the use of a split barge for container installation requires that the top of a geotextile container be located several metres below the water line.

6.5.6 Stability in currents over the structure

In [19] the stability of geotextile containers under current loads was modelled at a 1:20 scale. The results apply to currents perpendicular to the axis of a stack of geotextile containers. Four situations were modelled:

situation 1:
6 m high 4–3–2 stacking with crest at 0.56 m under water:
instability at a flow velocity of 0.57 m/s in front of the structure;

situation 2:
5 m high 4–3–2 stacking with crest at 3.50 m under water:
instability at a flow velocity of 1.42 m/s in front of the structure;

situation 3:
5 m high 4–3–2 stacking with crest at 4.70 m under water:
instability at a flow velocity of 1.70 m/s in front of the structure;

situation 4:
5 m high 3–2–1 stacking with crest at 3.50 m under water:
instability at a flow velocity of 1.34 m/s in front of the structure.

For the critical flow velocity across the crest of the stacked geotextile containers a design formula is given in [22, see paragraph 5.4.11]:

$$\frac{u_{cr}}{\sqrt{g \cdot \Delta_t \cdot D_k}} < 0.5 \text{ to } 1.0 \tag{6.15}$$

where:
u_{cr} = critical flow velocity across the crest of the structure [m/s];
g = acceleration due to gravity [m/s²].

Currents parallel to the axis of stacked geotextile containers allow much higher flow velocities. If the stacking is constructed of geotextile containers of a certain length in the sea or in a river or an estuary, then the current will veer from the structure. To determine the stability, an assumption has to be made for the flow velocity component perpendicular to the stacked containers upstream and/or above the crest (see figure 6.10). The upstream values can be comparable to the critical values from [19], with the values above the crest comparable to formula (6.15).

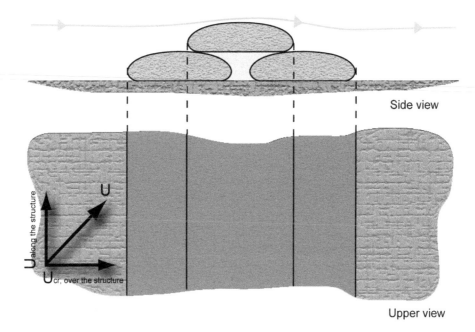

Figure 6.10 Definition sketch of critical flow velocity over and along the structure.

6.5.7   Stacking stability

Geotextile containers can be stacked steeply into a bund (1(V):2(H) to 1(V):3(H)). To build such a steep bund with containers special positioning techniques will be required.

During wave attack, a pressure differential occurs between the water pressure in the stacked containers and the external water pressure which may cause part of the stacked containers to shear off and thus make instability a threat. Currently, there is a limited amount of knowledge available relating to the stability of a stacked structure constructed of geotextile containers under different hydraulic loads. The Dutch Public Works Department has developed design criteria where a very conservative assumption is made on the distribution of water pressure that causes container shearing to be of concern. However, field tests suggest that the distribution of water pressure is less extreme and thus the stability greater. According to [15] container shearing of is the most critical aspect for stacked containers up to approximately the still water line. Container shearing can be a limiting factor with regard to instability of the uppermost container (see 6.5.5) when the slope is steep and the container is large in relation to the wave height. Also, the stability of the foundation can be important, e.g. a soft clay foundation can deform considerably when a geotextile container is dropped on it and can be susceptible to plastic deformations (squeezing) and excessive settlements when a stack of containers is placed on top. This must be assessed during the design stage.

The design formulae for the stability of a stack, presented in this section, are mostly based on a limited number of tests and therefore must be used with the requisite caution. From tests on tubes (which are more numerous than tests on containers) it is known that the way the tubes are stacked and the position with respect to the water line has an influence. The way containers are stacked is not taken into account in the formulae presented here. Furthermore, it will be shown later that the placing accuracy of geotextile containers is limited, and this may lead to instability at lower wave heights than according to the formulae presented below.

Small-scale model tests have studied the wave conditions under which instability will occur [15]. The study used normally filled geotextile containers (46%) and, additionally, filled geotextile containers (70%) on a scale of 1:20. The slope of the stacked containers was 1(V):3.1(H) and 1(V):1.7(H) respectively.

The pressure difference over the outermost layer of geotextile containers on the seaward side appeared to be mainly dependent on the wave height and the water level in relation to the stacked crest. The critical condition occurs when there is a wave trough on the seaward side of the stack. At this point there is low water pressure on the seaward side and high water pressure on the landward side of the stack. This results in an outward water pressure gradient in the container stack which is greater the narrower the width of the stack. Based on the test results, the following empirical relationships have been established for maximum hydraulic heads over the outermost layer of the geotextile containers:

$$\frac{\Phi}{H_s} = 0.24 \cdot \ln\left(\frac{D_t}{B_{tot}} + 0.04\right) + 0.77 \quad \text{if } b/D_k \cong 6 \tag{6.16}$$

$$\frac{\Phi}{H_s} = 0.31 \cdot \ln\left(\frac{D_t}{B_{tot}} + 0.04\right) + 1.00 \quad \text{if } b/D_k \cong 3.5 \tag{6.17}$$

where:

Φ = expected maximum hydraulic head difference over the outermost layer of the geotextile containers [m];

D_t = thickness of the shear-susceptible layer of the geotextile containers – see Figure 6.11 [m];

B_{tot} = total width of the layer of the geotextile containers under consideration – see Figure 6.11 [m].

To establish stability requirements for container stacking, the calculation model of Nurmohamed [2, in Dutch] is used. For the angle under which the geotextile containers can shear (see Figure 6.12):

$$\beta = \alpha - \psi \tag{6.18}$$

where:

β = shearing angle of the outer geotextile containers [deg];

α = stack slope [deg];

ψ = fill dilatancy angle [deg].

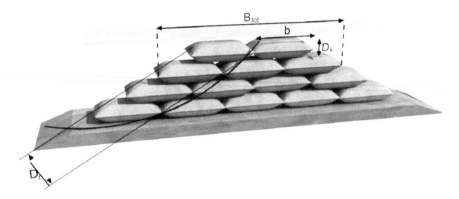

Figure 6.11 Definition sketch of container stacking.

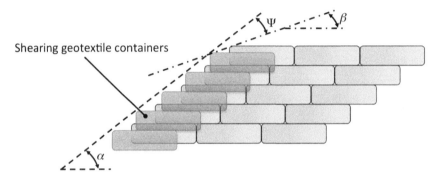

Figure 6.12 Definition sketch of container stacking with a predefined slip plane.

The case of $\beta = 0$ applies to the mobilised surface friction coefficient between horizontal geotextile containers:

$$f_m = \frac{F_0 \cdot \sin \alpha}{G - F_0 \cdot \cos \alpha} \tag{6.19}$$

where:

$$F_o = \frac{\rho_w \cdot g \cdot \Phi \cdot h_{stack}}{\sin \alpha} \tag{6.20}$$

where:

$\quad f_m$ = mobilised surface friction coefficient between adjacent geotextile containers [–];

$\quad F_o$ = outward force created by the hydraulic head assumed perpendicular to the slope per unit length [N/m];

G = submerged weight of the layer of geotextile containers per unit length [N/m];
h_{stack} = height of the stack [m].

The value of the surface friction coefficient between the adjacent geotextile containers depends on the type of geotextile used and can be tested when considered critical. For woven polypropylene geotextiles used in the geotextile container it may be approximated to:

$$f < \tan 30° \text{ to } \tan 35°$$

$$(6.21)$$

6.5.8 Influence of wave-induced liquefaction

Loosely placed sand has a tendency to compact under cyclic loading. When the pores are saturated with water, as is the case with geotextile containers under water, any water is expelled from the pores during compression. The pore-water flow necessary for this to occur requires a differential between the pore-water pressure inside and outside the geotextile container. When a cyclic load is applied to geotextile containers with loosely packed, saturated sand, excess pore pressures are created. In certain conditions these excess pore pressures can equal the weight of the upper-lying sand, causing liquefaction.

After a certain time the pore-water has dissipated, the sand will partially compact, giving rise to reduced susceptibility to further compaction and excess hydrostatic pressures. In time, the sand becomes so tightly packed that no excess hydrostatic pressure is generated any longer.

In section 6.3 of [16] a method is described to determine the liquefaction susceptibility of the sand in a geotextile container lying on the sea bed when it is subject to waves. According to this method there is no liquefaction problem if the wave period T is 3 s to 12 s, if the wave height gradually increases during a period of several hours and if:

$$T_d = \frac{d^2}{c_v} = \frac{\rho_w \cdot g \cdot d^2 \cdot \alpha_c}{k_s} \ll 300 \text{ s} \tag{6.22}$$

$$T_n = \frac{\Delta_n \cdot n}{\left((1-n) \cdot \alpha_c \cdot \psi_0 \right)} \ll 3000 \text{ s} \tag{6.23}$$

where:
T_d = characteristic drainage period [s];
T_n = characteristic compression period [s];
d = drainage distance ≈ height of the container [m];
c_v = consolidation coefficient for the grain skeleton [m²/s];
k_s = permeability of the fill material (sand) [m/s];
n = porosity of the fill material (sand) [–];
Δ_n = porosity reduction under constant wave load [–];
α_c = one-dimensional compressibility of the grain skeleton upon discharge [m²/kN];
ψ_0 = generation of excess pore pressure for undrained load [N/m² s].

The following can be assumed for geotextile containers with a cross-section of 10 m² when the wave period $T = 3$ s and when using sand with 20% particles finer than 63 µm, thus $D_{20} = 63$ µm:

Given:

drainage distance: $d = 2$ m;
permeability of the fill material: $k = 10^{-5}$ m/s;
porosity: $n = 0.4$;
porosity reduction under constant wave load: $\Delta_n = 0.01$;
one-dimensional compressibility of the grain skeleton during discharge: $\alpha_c = 3 \cdot 10^{-8}$ m²/N;
generation of pore-water pressure upon undrained load: ψ_0 = mean initial effective stress divided by the time needed for a number of waves to cause complete liquefaction in undrained conditions $\approx (\rho - \rho_w) \cdot g \cdot d/(30 \cdot T) \approx 20000$ N/m²/ $(30 \cdot 3$ s$) \approx 200$ N/m² s.

Substituting these values in formulae 6.22 and 6.23 gives $T_d = 117$ s and $T_n = 2777$ s, which indicates that complete liquefaction is unlikely, but that significant excess pore pressures may occur in this situation.

6.5.9   Placement accuracy

Before geotextile containers reach the bottom they may travel laterally (known as 'sailing'). Since this phenomenon can be explained easily in terms of flow mechanics, it is possible to simulate this with a mathematical model. The geotextile container does not fall straight downwards due to an initial oblique angle of the split barge or through currents or waves. During the fall the container is subjected to a horizontal force and moment, causing the container to move sideways. The extent of this movement (mean and standard deviation) increases with water depth, flow velocity and wave height. Currents tend to have more influence than waves. On an uneven bottom, for instance, when dropped on other geotextile containers the geotextile container comes to rest more quickly; on a flat bottom it takes a little longer and the potential horizontal shift is greater. Accuracy of placement is enhanced by a larger container mass and faster opening of the split barge.

Model tests were performed in the model facilities of Deltares (see Figure 6.13) as well as a prototype field demonstration [6]. These tests have provided insight into the drop process. Although uncertainty remains about the placement accuracy of geotextile containers dropped from a split barge, especially where this occurs at a water depth of more than 15 m. At a water depth of less than 10 m good placement accuracy can be achieved when dropping geotextile containers of approximately 300 m³ under the influence of certain currents and waves in order to build a stacked container structure with a side slope of 1(V):2(H).

At a water depth of more than 10 m the placement accuracy in currents and waves decreases markedly and it is possible that the centre of gravity of the container may move several metres horizontally during the fall.

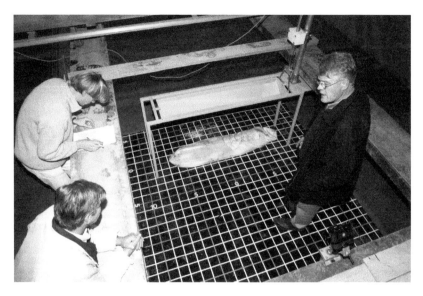

Figure 6.13 Model tests at Deltares on the dropping of geotextile containers.

To be able to place geotextile containers as accurately as possible, a thorough drop plan must be established. In view of the increasing difficulty of accurate placement at greater water depth, it is this area that demands most attention.

In constructing a stacked container structure it has to be accepted that a number of units may fall outside the profile. The first layer of containers is the most difficult to place ('sailing effect'). Taking account of the currents and waves a position will have to be established from where the first container is dropped. At substantial water depth the bed width of the profile of the structure to be built also increases. This widens the drop space for the first container (see Figure 6.14). After the first container is dropped, its resting position must be measured immediately. The position of the split barge during the drop and the position of the first container on the bed will indicate the placement accuracy. Following this, the second unit can be placed. After each drop the unit must be checked for its exact position since the placement accuracy is random in nature. During the drop of the initial layer, the containers to be dropped will be increasingly influenced by those already dropped, which helps the placement accuracy. For the second and subsequent layers the placement accuracy increases significantly, as stated previously. Field measurements and experiments have shown that the placement of the second and subsequent layers, despite the presence of currents and waves, can be twice or even four times as accurate as the dropping of the initial layer.

In summary, it can be concluded that placement accuracy increases under the following conditions:

- the lower the water depth;
- the lower the wave height and/or flow velocity;
- the greater the cross-section of the geotextile container and split barge;
- the greater the density of the geotextile container;

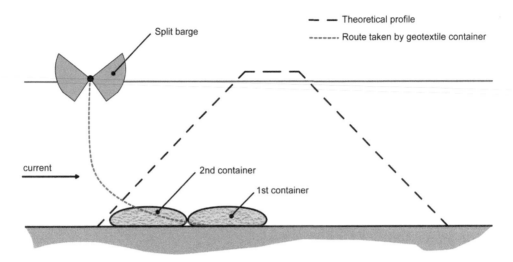

Figure 6.14 Definition sketch of the drop of geotextile containers.

- the more evenly the geotextile container is filled (with homogenous sand, in terms of grain-size distribution and water content, done layer by layer);
- the faster the split barge is opened;
- the drop (as for granular material) at sea occurs during a current tide window of less than or equal to 0.5 m/s;
- the rougher the surface of the bottom.

For further description and conclusions of the small-scale tests carried out at Deltares see Appendix H. For the small-scale tests a guidance system was developed in-house to demonstrate that large geotextile containers can also be accurately placed at significant depth. For the geotextile containers discussed in this chapter, using such a guidance system enables highly accurate placements to be achieved. Also, the loads generated during placement would be considerably lower.

6.6   CONSTRUCTION ASPECTS

The following points are relevant to construction:

- Within the entire construction process, filling is the time-determining factor and must be performed as quickly as possible. However, the container must be evenly filled. Dry filling is preferred to hydraulic filling;
- Dry filling allows steeper filling, which results in less width for the same filling volume and allows the geotextile container to more easily exit the split barge during the drop;
- For deeper drops (>8 m) "clean" sand is recommended for the fill material. By internal friction clean sand can absorb a major part of the fall energy that is generated during impact on the bottom. This significantly reduces the likelihood of geotextile rupture;

- For the drop of geotextile containers it is desirable for the split barge to fully open (50% of the width of the bin) within 20 to 30 seconds;
- The use of unsaturated sand (which contains air enclosed in the pores) as fill material is preferable because it temporarily provides more resistance to the deformation associated with impact and thus decreases the loads on the geotextile. Further, unsaturated sand has the advantage that the submerged weight of the geotextile container is lower and thus the fall velocity and fall energy are less than when saturated sand is used;
- To be able to place the geotextile containers as accurately as possible, a thorough drop plan must be drawn up. In view of the increasing difficulty of placement accuracy at greater water depths, it is this area that demands most attention;
- Geotextile containers should be dropped within the tide window. This increases the placement accuracy.

6.7   CALCULATION EXAMPLE

Situation

The case concerns a breakwater that will be constructed using geotextile containers. The containers will be dropped from a split barge. The following is known:

- maximum water depth is 10 m;
- length of the bin of the split barge: $l = 29.5$ m;
- maximum opening width of the split barge: $b_0 = 3$ m;
- filled cross sectional area in the horizontal plane: $A_s = 100$ m^2;
- fill material: sand;
- porosity of the fill material: $n = 0.4$ [–];
- geotextile: woven polypropylene (PP), maximum allowable tensile load in the geotextile $T_m = 140$ kN/m;
- geotextile: tensile stiffness: $J = 1000$ kN/m;
- drag force coefficient of the geotextile container: $C_d = 1.0$ [–];
- container dropped when there are no currents and waves;
- significant wave height $H_s = 3$ m;
- current velocity $u = 0.8$ m/s.

Required

Determine the required tensile strength of the geotextile for the various load situations. Assess the stability in waves and currents, the stability of the container stacking and the placement accuracy.

Calculation

Shape and dimensions of the geotextile container

Based on Table 6.1 and the split barge bin length of 29.5 m, the following dimensions of the geotextile container are used:

- degree of filling of the bin: $f = 0.8$ (80%);
- container volume: $V = 320$ m^3;

- cross-sectional area of the fill material in the bin: $A = 10.8$ m^2;
- circumference: $S = 18.3$ m;
- width: $b = 8.4$ m;
- height: $D_k = 1.6$ m.

For the fill density of the unsaturated geotextile container:

$$\rho = (1-n) \cdot \rho_s = (1-0.4) \cdot 2650 = 1590 \text{ kg/m}^3$$

For the fill density of the geotextile container fully saturated with water:

$$\rho = (1-n) \cdot \rho_s + n \cdot \rho_w = (1-0.4) \cdot 2650 + 0.4 \cdot 1000 = 1990 \text{ kg/m}^3$$

For the relative density of the geotextile container the formula (3.5):

$$\Delta_t = (1-n) \cdot \frac{\rho_s - \rho_w}{\rho_w} = (1-0.4) \cdot \frac{2650-1000}{1000} = 0.99$$

Required circumference

According to formula (6.2):

$$S \geq 2.5 \cdot \left(\frac{A}{b_0} + b_0 \right) = 2.5 \cdot \left(\frac{10.8}{3} + 3 \right) = 16.50 \text{ m}$$

$18.3 \geq 16.50 \Rightarrow$ Complies

Maximum tensile load acting on the geotextile during the opening of the split barge

According to formula (6.3):

$$T = 0.45 \cdot (\rho - \rho_w) \, g \cdot A$$
$$T = 0.45 \cdot (1990 - 1000) \cdot 9.81 \cdot 10.8 = 47.2 \text{ kN/m}$$

Maximum fall velocity of the geotextile container during free descent through the water

According to formula (6.9) the terminal fall (maximum possible) velocity is:

$$v_\infty = \sqrt{\frac{2 \cdot V}{A_s \cdot C_d} \cdot \frac{\rho - \rho_w}{\rho_w} \cdot g} = \sqrt{\frac{2 \cdot 320}{100 \cdot 1.0} \cdot \frac{1990 - 1000}{1000}} \cdot 9.81 = 7.9 \text{ m/s}$$

Tensile load upon impact

Initially it has been assumed that all of the fall energy can be absorbed by the geotextile alone, i.e. it will survive the impact irrespective of the fill material. If this is not the case, and in many cases this is not possible, the influence of the filling of the geotextile container will have to be taken into account as well. Without the influence of the fill material:

$$E_{val} = E_{geo}$$

With the formulae (6.5) and (6.8) and

$$T = J \cdot \varepsilon$$

The tensile load can be calculated on the basis of the fall velocity v_∞ (See Appendix G):

$$T = v_\infty \cdot \sqrt{\frac{A \cdot \rho \cdot J}{S}} = 7.9 \cdot \sqrt{\frac{10.8 \cdot 1990}{18.3}} = 270 \text{ kN/m}$$

Note: for the dimensions to be correct the calculation must be made with ρ in [kg/m³] and the geotextile tensile stiffness load to be resisted (270 kN/m). In this example this is (much) greater than the allowable tensile load of the geotextile (140 kN/m).

Since the fall velocity has a large influence on the result, it is better to determine the fall velocity more accurately using the formulae given in Appendix G where the fall velocity just before impact on the bed is calculated as 6.7 m/s, making the maximum tensile load = 223 kN/m. This is still too high for the geotextile, but the energy is also dissipated by other mechanisms, as described below.

Fall energy

It can also be assumed that the kinetic energy of the falling geotextile container is not only absorbed by the geotextile but in part also by the fill material (sand). For this situation formula (6.8) applies:

$$E_{fall} = \frac{1}{2} \cdot \rho \cdot A \cdot v^2 = 0.5 \cdot 1990 \cdot 10.8 \cdot 6.7^2 = 477.5 \text{ kJ/m} = E_{fill} + E_{geo}$$

Energy dissipation by the sand in the container

When the sand in the geotextile container is unsaturated, the effective stress in the sand will increase during the fall (the previously cited 'Bezuijen effect'). The extent of this increase depends on the level of saturation and the permeability of the material. In this example it is assumed that 80% of the increased pressure in the water (100 KPa) results in an increase in effective stress of: $\sigma'_i = 0.8 \times 100 = 80 \text{ kN/m}^2$. Substituting in formula (6.7):

$$E_{fill} = 2 \cdot \sigma'_i \cdot \tan\varphi \cdot \sqrt{A} \cdot (\sqrt{A} - 2) = 2 \cdot 80 \cdot \tan(30) \cdot \sqrt{A} \cdot (\sqrt{A} - 2) = 390.5 \text{ kJ/m}$$

The energy that has to be dissipated by the geotextile is given in formula (6.4):

$$E_{geo} = E_{fall} - E_{fill} = 477.5 - 390.5 = 90.0 \text{ kJ/m}$$

Formula (6.5) can now be used to calculate the strain in the geotextile:

$$\varepsilon = \sqrt{\frac{E_{geo}}{0.5 \cdot S \cdot J'}} = \sqrt{\frac{90}{0.5 \cdot 18.3 \cdot 1000}} = 10\%$$

The result is still a significant strain percentage but the corresponding tensile load is 100 kN/m, so a safety factor of 1.4 suffices compared the allowable load of 140 kN/m. In addition, there will be a reduction in the strength (80%) due to the seams. In this case $140 \cdot 0.8 = 112$ kN/m. This results in a safety factor of 1.12.

Stability in waves

For geotextile containers on the crest formula (6.14) applies:

$$\frac{H_s}{\Delta_t \cdot D_k} \leq 1 \Rightarrow H_s \leq \Delta_t \cdot D_k \Rightarrow H_s \leq 0.99 \cdot 1.6 = 1.58 \text{ m}$$

Given the calculated allowable significant wave height, different dimensions must be chosen for these containers. In many cases no container will be able to be used for the uppermost units of a structure (because containers cannot be dropped from a barge at the surface), so a different type of unit will be needed in this location (like a geotextile tube).

For the geotextile containers at a minimum of a single wave height below the still water line, the formula (6.14) again applies:

$$\frac{H_s}{\Delta_t \cdot D_k} \leq 2 \Rightarrow H_s \leq \Delta_t \cdot D_k \Rightarrow H_s \leq 2 \cdot 0.99 \cdot 1.6 = 3.17 \text{ m}$$

With regard to internal sand migration (caterpillar mechanism [37]) and the given wave height ($H_s = 3$ m) the crest should be at $2 \cdot H_s$ or more under water. In this case it is $2 \cdot 3 = 6$ m. The upper part of the construction could be built up using other types of units, such as geotextile tubes.

Stability in currents

According to formula (6.15):

$$\frac{u_{cr}}{\sqrt{g \cdot \Delta_t \cdot D_k}} < 0.5 \text{ to } 1.0$$

$$u_{cr} \leq 0.5 \text{ to } 1.0 \cdot \sqrt{g \cdot \Delta_t \cdot D_k} \Rightarrow u_{cr} \leq 0.5 \text{ to } 1.0 \cdot \sqrt{9.81 \cdot 0.99 \cdot 1.6} = 1.97 \text{ to } 3.94 \text{ m/s}$$

Stability of stacking

- For the entire (assumed) horizontal width of the stacking, three times the width of a geotextile container is maintained: $B_{tot} = 3 \cdot b = 25.2$ m;
- For the height of the shear susceptible layer of geotextile containers, $D_t = b \cdot \sin(\alpha) = 2.7$ m must be maintained;
- $H_s = 3$ m will be evaluated;
- $b/D_{car} = 5.25$. Choose formula (6.17) for safety purposes:

$$\frac{\Phi}{H_s} = 0.31 \cdot \ln\left(\frac{D_t}{B_{tot}} + 0.04\right) + 1.00 \Rightarrow \frac{\Phi}{H_s} = 0.31 \cdot \ln\left(\frac{2.7}{25.2} + 0.04\right) + 1.00 = 0.40$$

$$\Phi = 0.4 \cdot H_s = 1.2 \text{ m}$$

According to formula (6.20):

$$F_o = \frac{\rho_w \cdot g \cdot \Phi \cdot h_{st}}{\sin \alpha} = \frac{1000 \cdot 9.81 \cdot 1.2 \cdot 8}{\sin(18.43)} = 300 \text{ kN/m}$$

To determine the submerged weight G, the entire layer of containers has to be considered: the total height of the stacking $h_{st} = 10$ m formed by the six stacked geotextile containers (each container is proximally 1.6 m in height).

The downward force G of the containers under water per metre run can now be found using:

$$G = (1 - n) \cdot (\rho_s - \rho_w) \cdot g \cdot A \cdot \text{number of containers}$$
$$= (1 - 0.4) \cdot (2650 - 1000) \cdot 9.81 \cdot 10.8 \cdot 6 = 629 \text{ kN/m}$$

For the stability of the stacking formula (6.19) applies:

$$f_m = \frac{F_0 \cdot \sin \alpha}{G - F_0 \cdot \cos \alpha} = \frac{300 \cdot \sin(18.43)}{629 - 300 \cdot \cos(18.43)} = 0.275$$

According to formula (6.21):

$$f < \tan 30° \text{ to } \tan 35° = 0.58 \text{ to } 0.70 \Rightarrow \text{complies}$$

Placement accuracy at 10 m water depth

The placement accuracy can be determined in two ways (see Appendix H):

1. $s_p = 0.4 \cdot h - 3.2 = 0.8$ m
2. $s_p = c \cdot \sqrt{h \cdot D_{50}} = 0.7 \cdot \sqrt{10 \cdot 1.6} = 2.80$ m

The $s_p = 0.8$ m is based on the formula derived from measurements obtained at Kandia dam during the placement of geotextile containers. If the drop area under consideration and the geotextile containers to be used largely corresponds with the situation and units used during the building of the Kandia dam, then use of this formula is recommended to approximate the placement accuracy.

The second formula is derived from the drop of stones. Theoretically, this should correspond to the placing accuracy during the drop of geotextile containers. Since only a single type of geotextile container is almost always dropped at one location, it is not possible to check this authenticity of this formula with practice. The second formula is therefore only recommended for use when determining the scope of situations that deviate in relation to the first formula.

It should be noted that the calculated placement accuracies are only estimates and must not be considered actual. They are purely indicative. In practice, measurements must always be taken to establish the placement accuracy in any installation.

Appendix A

Permeability of geotextiles

The permeability of geotextiles in general, and thus also of geotextile-encapsulated sand elements, has to be great enough to prevent excessive pore pressures from developing. A safe assumption is that the permeability of the geotextile must be at least 10 times that of the permeability of the fill material it is filtering. Since geotextile permeability is difficult to establish, it can also be proposed that there must be a minimum water pressure differential across the geotextile. For the case of laminar flow, the hydraulic conductivity of the fill material is described by Darcy's Law:

$$q = k_s \cdot i \tag{A.1}$$

where:

q = specific discharge (=Q/A_g) [m/s];
k_s = hydraulic conductivity of the fill material [m/s];
i = hydraulic gradient in the fill material [–].

Table A.1 lists typical hydraulic conductivity values for various granular fill materials.

The flow through a geotextile with sand on one or both sides is usually laminar since the flow velocity through the geotextile is controlled by the adjacent sand and so Darcy's Law, based on laminar water flows, can be used in determining the permeability

Table A.1 Hydraulic conductivity of various fill materials [24].

Material	D_{50} [m]	k_s [m/s]	Type of internal water movement
Clay	$<2.10^{-6}$	10^{-10}–10^{-8}	Laminar
Silt	2.10^{-6}–63.10^{-6}	10^{-8}–10^{-6}	Laminar
Sand	63.10^{-6}–2.10^{-3}	10^{-6}–10^{-3}	Laminar
Gravel	2.10^{-3}–63.10^{-3}	10^{-3}–10^{-1}	Turbulent
Armourstone	63.10^{-3}–0.4	10^{-1}–5.10^{-1}	Turbulent
Armourstone (coarse)	0.4–1	5.10^{-1}–1	Turbulent

of a geotextile in the same manner as it is done for the fill material. However, in practice, it is common to express the permeability of a geotextile as permittivity:

$$\Psi = \frac{q}{\Delta H} \tag{A.2}$$

where:

Ψ = permittivity [s^{-1}];
q = specific discharge [m/s];
ΔH = water head differential across the geotextile [m].

The permittivity is the ratio between the specific discharge and the water head differential across the geotextile. By using permittivity, geotextile thickness is not required to be known. Conversely, when using geotextile permeability, geotextile thickness is required to be known and this can give rise to significant errors because of the inaccuracy in measuring geotextile thickness.

In Europe, geotextiles are supplied according to CEN standard methods of test (the European Norms that govern geotextile methods of test) by the manufacturer. The CEN method of test for geotextile water flow is based on the discharge through the geotextile at a constant water head differential of 50 mm which results in the determination of a "velocity index" parameter. While this velocity index parameter has the units m/s it should not be confused with the geotextile permeability value, which also has the units m/s. However, the velocity index can be used to determine both the permittivity and the permeability of the geotextile.

For permittivity, the following formula (A.3) applies:

$$\Psi = \frac{q}{\Delta H} = 20 \cdot V_{H50} \tag{A.3}$$

Where V_{H50} is the velocity index determined in accordance with EN ISO 11058. This relationship is only valid if V_{H50} is given in m/s. To calculate the permeability, the thickness of the geotextile must also be known. Table A.2 gives typical permittivity values for different geotextiles.

The permittivity of much of the geotextiles used for geotextile-encapsulated sand elements (woven polypropylene) is in the range of 0.1–0.6 s^{-1} at a specific discharge of 0.01 m/s. The openings of geotextiles, and thus the permittivity, is limited by the sand density requirement.

Table A.2 Permittivity values for various geotextiles [24].

Type of geotextile	O_{90} [mm]	Ψ [s^{-1}]
Monofilament	0.1–1	1–5
Tape yarns	0.05–0.6	0.1–1
Woven	0.2–1	0.05–0.5
Nonwoven	0.02–0.2	0.01–2

At the transition between the fill material and geotextile the specific discharge is constant and thus:

$$\Psi \cdot \Delta H = k_s \cdot i \tag{A.4}$$

where k_s is the hydraulic conductivity of the soil [m/s] directly adjacent to the geotextile.

No excess pore pressure will be created if the pressure fall across the geotextile is sufficiently small. A safe assumption is therefore:

$$\Delta H = \frac{k_s \cdot i}{\Psi} \leq 0.01 \text{ m} \tag{A.5}$$

The permeability of a geotextile will vary as a result of its interaction with the adjacent fill material. When adjacent fill material is present, this interaction will always result in a lower measured geotextile permeability than if the geotextile was tested in-isolation. When adjacent fill is present three phenomena may occur, namely, blinding, blocking and clogging.

Blinding occurs when the fill material has grains that are large compared to the openings in the geotextile. An example is a geotextile that is blinded by the presence of adjacent armour stone. Here, the armour stone lies on the geotextile and no water can flow through the geotextile (where this blinding occurs) so the average permeability of the geotextile is lower.

If the particles of the fill material are moved by the flow of water and deposited in the openings of the geotextile blocking occurs, a phenomenon that can occur in thin materials and exists when the following factors are present at the same time:

- a uniform opening size exists in the geotextile;
- a uniform grain-size distribution exists in the fill material;
- O_{90}/D_{90} lies between 0.5 and 1.0.

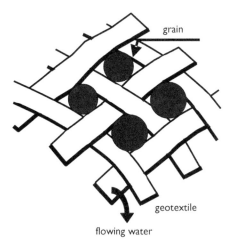

Figure A.1 Blocking mechanism.

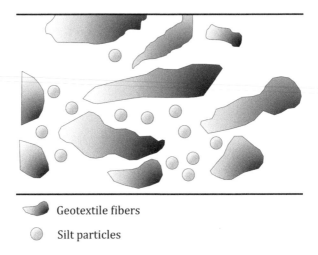

Geotextile fibers

Silt particles

Figure A.2 Clogging mechanism.

With clogging, (see Figure A.2) the geotextile pores fill up because fine particles of the fill material force their way into the geotextile structure and block the pore channels. This has also been called 'mineral clogging' and occurs mostly in nonwoven geotextiles. Despite clogging reducing the permeability of geotextiles by some 50%, studies have shown that the resulting permeability of nonwoven geotextiles is still greater than that of the fill material. It is proposed [22, paragraph 4.5.3.1] that clogging will not occur if:

- For $C_u > 3$: $O_{95}/D_{b15} > 3$;
- For $C_u < 3$: criterion for internal stability of fill should be satisfied and/or a geotextile with maximum opening size from the soil-tightness criteria should be specified.

Blocking can lead to a reduction of the permittivity (and thus of the permeability too) by a factor of 5. Clogging can cause even a higher reduction to occur [39].

CALCULATION EXAMPLE

Situation

Geotextile containers have been selected as the filled elements for the construction of a dam. The containers will be filled with sand ($k_s = 10^{-4}$ m/s, $D_{90} = 0.2$ mm and $D_{15} = 0.1$ mm). In the vertical plane the container has a width of 7 m and height of 2 m. The water level is on one side of the container 0.5 m higher than on the other side.

The supplier prescribes a geotextile that is characterized by the following properties: $O_{90} = 0.6$ mm and a velocity index of 0.004 m/s determined in accordance with EN ISO 11058. Thus, using formula A.3 $\Psi = 0.08$ s^{-1}).

Requested

Check whether this geotextile complies with regard to excess pressure, blocking and clogging.

Calculation

Using formula (A.4):

$$k_s \cdot i_s = \Psi \cdot \Delta H$$

Now, the hydraulic gradient i_s in the geotextile container can be determined from (where B_s is the width of the geotextile container):

$$i_s = \frac{\Delta H}{B_s} = \frac{0.5}{7} = 0.071$$

With the geotextile properties as stated by the supplier, a check can be made on whether the excess pressure is low enough. It must comply with formula (A.5):

$$\Delta H = \frac{k_s \cdot i_s}{\Psi} \leq 0.01 \text{ m}$$

Substituting the various values in the above relationship gives:

$$\Delta H = \frac{k_s \cdot i_s}{\Psi} = \frac{10^{-4} \cdot 0.071}{0.08} = 8.9 \cdot 10^{-5} \text{ m} \implies \text{complies}$$

Blocking occurs if the following conditions are present at the same time:

- A uniform opening size of the geotextile;
- A uniform grain-size distribution of the fill material;
- $\frac{O_{90}}{D_{90}} = 0.5$ to $1.0 \implies \frac{0.6}{0.2} = 3 \implies$ complies.

Therefore, there is little possibility of blocking.
The last step is to check for clogging. This will not occur if:

$$\frac{O_{90}}{D_{15}} > 3 \implies \frac{0.6}{0.1} = 6 \implies \text{complies}$$

The check shows that the geotextile prescribed by the supplier complies and can be used.

CE marking

It is compulsory for all geotextiles used within the European Union to carry a CE Marking, e.g. see Figure B.1. In the context of this manual not every detail of the CE marking can be dealt with here, so only those aspects relating to the user of geotextiles are covered.

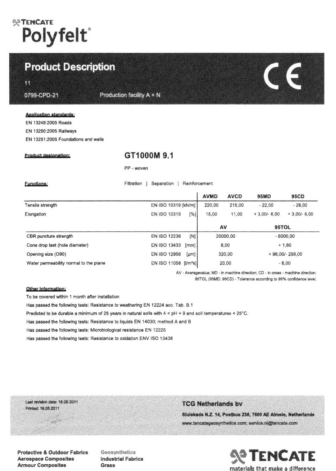

Figure B.1 Example of CE marking.

For geotextiles there is no classification within the CE marking and so it cannot be established that a geotextile must have, for example, a tensile strength greater than 30 kN/m for use in a hydraulic application. CE marking distinguishes various functions for which a geotextile can be used. They are as follows:

- reinforcement;
- separation;
- filtration;
- drainage.

To be able to assess the extent to which a geotextile can fulfill the required function, prescriptions are made per application area on which (characteristic) properties of the geotextile must be made available by the manufacturer to be able to compare geotextiles with each other and, where possible, judge their performance in-situ. Minimum requirements are described in the application standards EN 13249 to EN 13257 and EN 13265 for geotextiles. Of importance for the application of geotextile-encapsulated sand elements is EN 13253 "Geotextiles and geotextile – related products – Required characteristics for use in erosion control works".

If a geotextile is considered suitable for filtration then the "water permeability normal to the plane", for example, must be known. This means that only a value measured according to the representative standard is known for the permeability, not that it must be greater than a certain minimum value, which is dependent on in-soil conditions. Three kinds of test have been defined:

- The so-called "harmonized tests", the results of which are used not only to determine a parameter but also to indicate whether the quality of the product is consistent. The manufacturer has to indicate which quality procedure is followed to ensure that the test results on his product remain within the spread indicated by the manufacturer (H).
- The tests that have to be performed in all cases. The results of the tests that are stated for a function must be known in order to use the geotextile for that function (A).
- The tests that are required in some cases only, such as certain chemical tests that only have to be performed if the geotextile is expected to come into contact with different environments (S).

An example of how the information is presented in a CE application standard is shown in Table B.1. The results from the stated tests are made available upon delivery of the geotextile to site. The CE marking (on the geotextile rolls) shows the properties necessary to meet the relevant application standard. The advantages for the user of geotextiles containing a CE marking are:

- the manufacturer has a recognized quality control system, along with an independent verification organization examining whether there is sufficient justification to present the values and their variation as stated by the manufacturer. According to the European regulations geosynthetics (which include geotextiles) are a category of products where the manufacturer can state what will be their value and their

Table B.1 Table from the European standard: Geotextiles and geotextile-related products–Characteristics required for use in erosion control works (coastal protection, bank revetments).

Characteristic	Test method	Functions		
		Filtration	Separation	Reinforcement
1 Tensile strength[b]	EN ISO 10319	H	H	H
2 Elongation at maximum load	EN ISO 10319	A	A	H
3 Tensile strength of seams and joints	EN ISO 10321	S	S	S
4 Static puncture (CBR test)[a,b]	EN ISO 12236	S	H	H
5 Dynamic perforation resistance (cone drop test)[a]	EN 918	H	A	H
6 Friction characteristics	prEN ISO 12957-1: 1997 and prEN ISO 12957-2: 1997	S	S	A
7 Tensile creep	EN ISO 13431	–	–	A
8 Damage during installation	ENV ISO 10722-1	A	A	A
9 Characteristic opening size	EN ISO 12956	H	A	–
10 Water permeability normal to the plane	EN ISO 11058	H	A	A
11 Durability	According to annex B	H	H	H
11.1 Resistance to weathering	ENV 12224	A	A	A
11.2 Resistance to chemical ageing	ENV ISO 12960 or ENV ISO 13438, ENV 12447	S	S	S
11.3 Resistance to microbiological degradation	ENV 12225	S	S	S

Relevancy:
H: required for harmonization.
A: relevant to all conditions of use.
S : relevant to specific conditions of use.
– : indicates that the characteristic is not relevant for that function.
[a]   it should be considered that this test may not be applicable for some types of products, e.g. geogrids.
[b]   if the mechanical properties (tensile strength and static puncture) are coded "*H*" in this table the producer shall provide data for both. The use of only one, either tensile strength or static puncture, is sufficient in the specification.

variation in tensile strength or whatever property of a harmonized test that is applicable for a specific function. These values are not controlled by an external body. External bodies only control the quality system of the manufacturer;

- the results of a specific number of index tests are defined and can be used with safety.

It should be noted that the results of index tests, as required for the CE marking, can only be used indirectly, for design index tests by nature do not take into consideration the interaction with the ground. Table B.1 shows which tests are necessary for the applications in this manual.

DURABILITY IN CE MARKING

The application standards stated above also contain a 'normative annex' on durability in which the following situations are differentiated:

* The extent to which the geotextile is resistant to ultraviolet (UV) radiation. This is tested in a "Weathering test" (ENV 12224) and is important to empirically determine how long a geotextile can be exposed to UV radiation before it is covered by soil or water;
* The durability after installation of the geotextile.

RESISTANCE TO UV RADIATION

The Weathering test (ENV 12224) exposes a geotextile to a predetermined amount of UV radiation with the strength established before and after the UV radiation period. Depending on the result of the test and the required application, the UV exposure times stated in Table B.2 reveal how long the unprotected geotextile may be exposed to sunlight. Geotextiles not tested have to be covered with soil within one day after placement, or placed deep enough in the water that they are no longer subject to UV light exposure.

The text above is taken from the durability appendix of the CEN standards. Geotextile mattresses, tubes, bags and containers can normally be exposed for a long period to sunlight. This is normally not the case for other systems, such like slope revetments, filter constructions under water, etc. If required, the working method can be adjusted to ensure that the geotextile is subjected to sunlight for only a short period of time.

DURABILITY AFTER INSTALLATION OF THE GEOTEXTILE

In the European standards for geotextiles two situations are distinguished:

* applications with a lifetime shorter than 5 years;
* applications with a lifetime of up to 25 years.

Table B.2 Maximum times that a geotextile may lie exposed on a construction site, depending on the result of the 'Weathering Test' and the application.

Application	Retained strength (after exposure to test)	Maximum time of exposure after installation
Reinforcement or other applications where long-term strength is a significant parameter	>80%	1 month[a]
	60% to 80%	2 weeks
	<60%	1 day
Other applications	>60%	1 month[a]
	20% to 60%	2 weeks
	<20%	1 day

[a] Exposure of up to 4 months may be acceptable depending on the season and on the location in Europe.

Table B.3 Basic materials for geotextiles and prescribed durability tests for up to 25 years design life.

Material	Test
Polyester (PET)	Hydrolysis, ENV 12447
Polypropylene (PP)	Oxidation test, ENV ISO 13438
Polyethylene (PE)	Oxidation test, ENV ISO 13438

The materials from which geotextiles are normally made, polyester (polyethylene terephthalate), polyethylene, and polypropylene, or combinations thereof, are sufficiently durable for structures with a lifetime shorter than five years and thus require no further testing. This applies to all applications, with the exception of sustainable reinforcement, on condition that the pH value (acidity level) of the soil lies between 4 and 9 and that no recycled geotextile is used.

For structures with a lifetime of up to 25 years, supplementary durability tests are required, see Table B.3. The product is suitable if more than 50% of the original strength is retained after completion of the test.

The European standard also allows the durability to be determined based on experiences with comparable products (both raw materials and processing) under similar circumstances. Furthermore, the durability annex of the standard describes ground situations with a high or low pH and other specific circumstances that are normally not relevant for geotextile-encapsulated sand elements.

LONGER LIFETIME

Structures in which geotextiles are used are often designed for a lifetime considerably exceeding 25 years and experience suggests that such structures are easily capable of longer design lives. That the standard does not go beyond 25 years does not mean that the structures cannot exceed this period but that the durability cannot be accurately assessed using these rather general procedures. Also of importance is the biological activity in the ground. This is an area where assessments can be made based on experience and where data and no general guidelines exist.

In practice, sometimes two extreme positions are taken of which both are incorrect. On the one hand the durability of a structure exceeding 25 years may be doubted, with the argument that such a period is not contained in the standards. This position is unjustified since experience of lifetimes greater than 30 years have revealed little geotextile ageing. On the other hand, a lifetime of 150 years (or more) is sometimes claimed on the basis of an oxidation test alone.

Today, some specialized laboratories can do durability tests that prove the lifetime of a geotextile can be longer than the lifetimes mentioned above. Their results show that by using specific polymer additives long lifetimes are possible.

Appendix C

Design charts for the geotechnical stability of geotextile mattresses and geotextile bags on slope revetments

A wave load on the top layer of a slope revetment continues on into the subsoil. The wave transmission is muted and delayed because the water in the pore space is elastically compressible. This causes fluctuating negative and positive pore (water) pressures in the subsoil and thus a corresponding increase and decrease in effective stress. This phenomenon is called elastic storage and can lead to shearing of the subsoil. The stability of the subsoil can become critical if the elastic storage causes a decrease in the effective stress to such an extent that the ratio between the shear stress and the normal stress exceeds $\tan(\phi)$ of the subsoil, where ϕ is the internal friction angle.

For the criterion of shearing of the subsoil design charts have been drawn up for the most common situations (see Figures C.1 and C.2). In these charts the maximum allowable wave height has been plotted against the thickness of the top layers, for different slope angles. The design charts are applicable to geotextile mattresses and

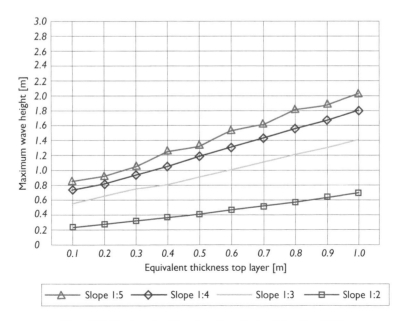

Figure C.1 Design chart for shearing of subsoil for $S_{op} = 0.03$.

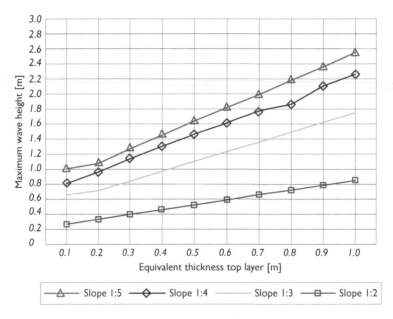

Figure C.2 Design chart for shearing of subsoil for $S_{op} = 0.05$.

bags that lie directly on a sandy subsoil and are shown for two values of wave steepness, for $s_{op} = 0.03$ (Figure C.1) and for $s_{op} = 0.05$ (Figure C.2).

The design charts [22] contain an equivalent top layer thickness. If the units are placed directly on sand, the equivalent thickness is equal to the actual thickness of the top layer, where the top layer must be determined as an average over the surface ($D_{eq} = D_k$).

Dimensions and shape of geotextile tubes

When a geotextile tube is empty and lying flat on the ground surface, its width is equal to half its circumference. When fully filled (degree of filling = 1.0), it has a circular shape with a radius R = circumference/2π). In practice, a degree of filling of between 0.60 and 0.85 can only be achieved. At this degree of filling a shape is obtained where the underside of the cross-section is flat, the sides approximate quadrants of a circle and the upper side approximates a (half) ellipse. The radius of this ellipse along the horizontal axis is equal to the radius of the quadrant (quarter of the circle). The filled dimensions of the geotextile tube can be approximated using the mathematical formulae given below, and using the basic parameters shown in Figure D.1.

Note: The best way to determine the shape of the geotextile tube is to use the method described by Timoshenko [44] and which is used in several computer programs. Bezuijen and Van Steeg [40] have shown that this formulation presents excellent agreement with calculations and measurements. However, this method requires the use of a computer and numerical calculations. When these are not available, it is possible to make an estimate of the tube shape using the method described below.

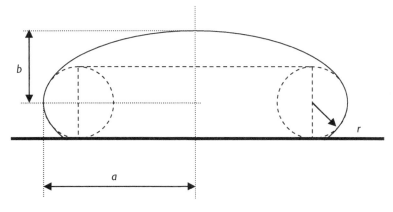

Figure D.1 Definition sketch of a filled geotextile tube.

TUBE CIRCUMFERENCE

The tube circumference (S), which is independent of the degree of filling, comprises the components of a rectangle, two circle quadrants plus half an ellipse. By setting this filled circumference equal to the circumference of a notional circle with 100% degree of filling and radius R, a formula with several unknowns arises.

$$S = \left(2 \cdot \alpha^2 + \pi - 2\right) \cdot r + M = 2 \cdot \pi \cdot R \qquad (D.1)$$

where:

$$\alpha = \frac{b}{r}$$

$$M = \pi \cdot \alpha^2 \cdot r \cdot \frac{\sqrt{1 + \dfrac{35}{72} \cdot m^2 + \dfrac{2}{15} \cdot m^4}}{1 + m}$$

$$m = \frac{a - b}{a + b} = \frac{\alpha - 1}{\alpha + 1}$$

where:
S = circumference of the tube [m];
M = half circumference (approximate) of a half ellipse [m];
b = half height of an ellipse [m];
a = half width of an ellipse [m];
r = radius of curvature of the ellipse along the horizontal axis = the radius of the quadrants [m];
R = the radius of the notional circle at 100% degree of filling [m].

TUBE CROSS-SECTIONAL AREA

The filled cross-sectional area of the tube depends on the degree of filling and comprises a half rectangle, two circle quadrants plus half an ellipse. By setting the calculated cross-sectional area equal to the product of the degree of filling and the cross-sectional area of the notional circle (at 100% filling) the following formula, with the same unknowns, is generated.

$$A = \frac{1}{2} \cdot \pi \cdot \alpha^3 \cdot r^2 + 2 \cdot \alpha^2 \cdot r^2 + \left(\frac{1}{2} \cdot \pi - 2\right) \cdot r^2 = f \cdot \pi \cdot R^2 \qquad (D.2)$$

where:
A = cross-sectional area [m^2];
f = degree of filling [–].

Table D.I Shape and dimensions of geotextile tubes for various degrees of filling. The shaded area represents the most common degree of filling, based on the cross-sectional area of the geotextile tube.

f [–]	r[m]	B [m]	h [m]
1.00	1.00 R	2.00 R	2.00 R
0.95	0.70 R	2.28 R	1.59 R
0.90	0.58 R	2.40 R	1.42 R
0.85	0.50 R	2.49 R	1.29 R
0.80	0.43 R	2.56 R	1.17 R
0.75	0.37 R	2.63 R	1.07 R
0.70	0.32 R	2.69 R	0.98 R
0.65	0.28 R	2.74 R	0.89 R
0.60	0.24 R	2.79 R	0.81 R

f = degree of filling.
r = radius of the quadrants for both sides of the straight basis of the geotextile tube.
B = the width of the geotextile tube [m].
h = height of the geotextile tube.
R = radius of the notional circle for 100% degree of filling.

TUBE WIDTH AND HEIGHT

If a specific degree of filling is chosen, then the width and the height of the geotextile tube can be expressed in terms of the radius of the notional circle for 100% degree of filling.

$$B = 2 \cdot \alpha^2 \cdot r$$
$$h = (1 + \alpha) \cdot r$$

where:
   B = width of the filled geotextile tube [m];
   h = height of the filled geotextile tube [m].

   Table D.1 lists degree of filling values along with comparative values of tube width and height in terms of R.

Shape and forces in geotextile tubes (Timoshenko method)

The method described by Timoshenko [44] and used in various computer programs to calculate the shape of filled tubes (e.g. GeoCoPS, SOFTWINN, Deltares) is based on the tensile stress being constant over the entire circumference of the geotextile, except for that part lying on the ground surface and with the material having no flexural stiffness. The fill material is considered to behave as a liquid with the pressure exerted by the fill material in the geotextile tube counteracted by the curvature and the tensile load in the geotextile. The curvature is determined by the local pressure in the fill material (see Figure E.1):

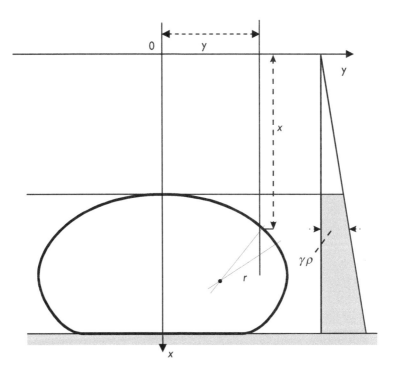

Figure E.1 Definition sketch for calculations using the Timoshenko method.

$$T = p \cdot r \tag{E.1}$$

where:
  T = tensile load in the geotextile [kN/m];
  p = pressure in the fill material [kPa];
  r = radius of curvature at a random point on the circumference of the geotextile tube [m].

The resulting formulae have analytical solutions but due to their complexity these can only be solved using numerical methods. Without using Table D.1 the calculation does not lend itself to a 'quick calculation by hand'.

The results, according to Timoshenko, can be shown graphically (see Figures 5.7 to 5.9). Figure E.2 shows one set of calculations using a 12 m circumference tube.

In practice, the hydrostatic pressure on the upper tube side lies within much narrower margins than shown in Figure E.2, normally between the 0.5 and 1 m water

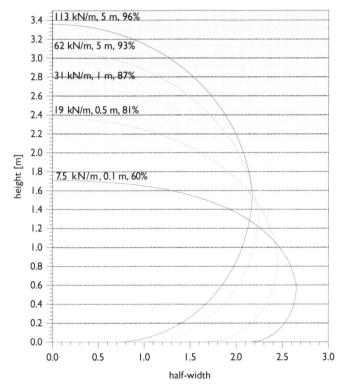

Figure E.2 Results of numerical calculations. Shape of a geotextile tube with a circumference of 12 m for various degree of filling values (based on volume). The numbers for each line shows respectively the tensile load in the geotextile, the pressure head in metres of water on the upper side of the tube and the degree of filling (expressed as a percentage). Calculations have been performed for a fill material that has a 10 kPa higher pressure than the external surroundings (e.g., filling with a sand-water mixture under water). Clearly the tensile load in the geotextile increases significantly at a higher degree of filling.

head (water pressure 5 to 10 kPa). Consequently, the degree of filling is theoretically between 81% and 87%, depending on the tube dimensions (larger tubes result in a lower degree of filling with the same filling pressure on top) and whether or not the filling is carried out above the water line (filling below the water line leads to a higher degree of filling). Since the sand-water mixture volume decreases after filling, the practical degree of filling is almost always less than 80%. A greater range has been incorporated here to demonstrate how the degree of filling and the excess pressure influence the shape of the geotextile tube and what its impact is on the maximum tensile load. Figure E.2 also shows that when the degree of filling is higher, through extra pressure applied during filling, this leads to an extra rise in the tensile load in the geotextile.

One disadvantage with this method is that the pressure in the fill material and the tensile load in the geotextile are used as input parameters with output in the form of the cross-section of the geotextile tube. In designing geotextile tubes it is usual to use the reverse approach where the circumference (or diameter) of the geotextile tube is used as input in order to determine the required tensile strength of the geotextile as output.

LESHCHINSKY (GEOCOPS)

Leshchinsky has translated the Timoshenko method into a computer program, GeoCoPS [43], which can calculate both the tensile load in the geotextile and the cross-sectional shape of the geotextile tube. Formulae for the tensile load in both the circumferential and axial direction of the geotextile tube are given below. For further details refer to [21].

For the tensile load in the circumferential direction around the geotextile tube (see Figure E.3):

$$T = \frac{(p_0 + \rho \cdot x) \cdot [1 + (y')^2]^{\frac{3}{2}}}{y''} \tag{E.2}$$

where: $y' = \dfrac{dy}{dx}$ and $y'' = \dfrac{d^2 y}{dx^2}$

where:

T = circumferential tensile load in the geotextile tube [kN/m];
p_0 = excess pressure (also known as pump pressure) [kPa];
ρ = fill density [kN/m³].

Because:

$$r = \frac{[1 + (y')^2]^{\frac{3}{2}}}{y''} \tag{E.3}$$

this formula is identical to formula (E.1) for $p = p_0 + \rho \cdot x$.

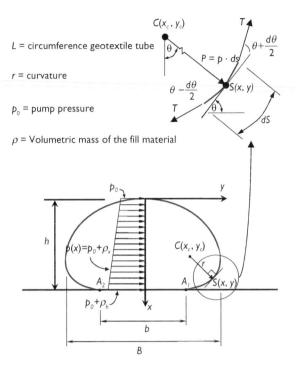

L = circumference geotextile tube

r = curvature

p_0 = pump pressure

ρ = Volumetric mass of the fill material

Figure E.3 Definition sketch of geotextile tube for calculations according to Leshchinsky.

In Figure E.3 the fill density ρ is stated as a parameter. This applies when the geotextile tube is exposed in air. When the geotextile tube lies in water the fill density parameter ρ is the buoyant fill density.

The tensile load in the axial (longitudinal) direction of the geotextile tube is also determined using the pressure of the fill material (see Figure E.4):

$$T_{axial} = \frac{2}{L} \cdot \int_0^k [p_0 + (\rho \cdot x)] \cdot y(x)dx \qquad (E.4)$$

If the shape of the geotextile tube is known and the pressure is known, this formula can be used to calculate the axial load.

The user can enter the required circumference of the geotextile tube and the pump pressure and the computer program calculates the corresponding shape of the geotextile tube (height, width and cross sectional area) and the required tensile strength of the geotextile to be used (based on the calculated tensile load).

SYLVESTER

In Sylvester [25] a design graph for geotextile tubes has been drawn up based on the results of a small-scale model study. Calculations have demonstrated that this method

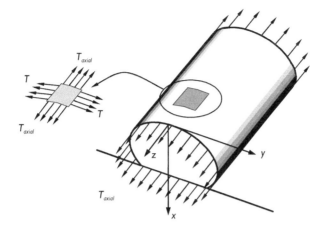

Figure E.4 Axial tensile load in the geotextile.

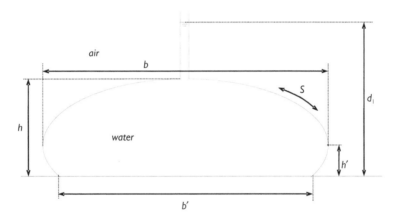

Figure E.5 Definition sketch of geotextile tube according to Sylvester.

generates similar results to GeoCoPS. Because Sylvester has shown the solution for different cases in a dimensionless chart, this method is explained in more detail below.

The method is based on small-scale model research with flexible geotextile tubes (plastic), filled with water, and performed at the University of Western Australia. Depending on the water pressure in the tube, and its circumference, different values were determined for both the shape of the geotextile tube (height, width and filled cross-sectional area) and the tensile load in the geotextile tube. From these results, relationships were derived between the input parameter ratio (water pressure (d_1)/ tube circumference (S)) and the following ratios (see also Figure E.5):

- height/width: h/b;
- width/circumference: b/S;
- filled cross sectional area/rectangle $b \cdot h$: $A/(b \cdot h)$;

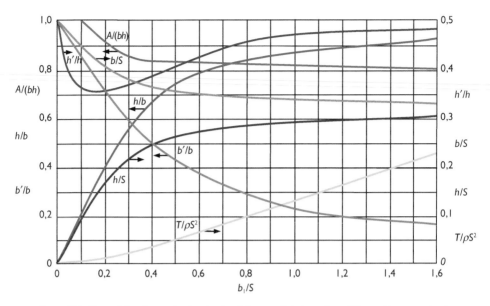

Figure E.6 Plot of the dimensionless ratios. The graph is based on filling with water in air.

- bed contact/total width: b'/b;
- height at maximum width/total height: h'/h;
- circumferential tensile load in the geotextile/$\rho \cdot S^2$: $T/\rho \cdot S^2$.

These dimensionless ratios are then combined into a design chart (see Figure E.6) that can be used for the design of the shape of geotextile tubes filled (under pressure) with a sand-water mixture. This design chart shows, among other things, that for $b_1/S > 0.4$ the filled cross-sectional area ratio ($A/(b \cdot h)$) of the geotextile tube remains relatively constant. Given that the tensile load in the geotextile increases significantly along the same trajectory (by 600%), it is recommended to limit $b_1/S \leq 0.4$, which results in a relatively flat geotextile tube.

Tensile load in geotextile container during the opening of the split barge

This appendix describes the overall methodology used to determine the tensile load generated in a geotextile container while opening the split barge, based on the theory described in [17] and [22]. For more detailed information these two references should be consulted.

The geotextile container goes through four stages during the opening of the split barge (see Figure F.1).

- I: stretching of the lower part of the geotextile during the beginning of the barge opening;
- II: shift of the container towards the barge opening without noticeable deformation;
- III: shift of the container through the barge opening with deformation of the lower part of the geotextile;
- IV: passage of the container through the barge opening with deformation of the entire geotextile.

STAGE I

To facilitate the dumping of the geotextile container during opening of the split barge, the geotextile is laid folded on the floor of the barge. When the barge is opened, the folded part drops through the opening. Subsequently, when increasing the opening, the geotextile becomes subject to tensile loads. The tensile loads are in equilibrium with the friction between the geotextile and the sides of the split barge. The suspended part, and the tensile load generated, are greater the further the split barge is opened. This continues until the tensile load is great enough so that the other parts of the geotextile begin to shift downwards. The maximum tensile load in the geotextile during this first stage of the opening of the split barge can be written as:

$$T = 0.5 \cdot \frac{W'}{L} \cdot \tan(\varphi - \theta) \qquad (F.1)$$

where:
T = tensile load in the geotextile [kN/m];
W' = submerged weight of the geotextile container [kN];

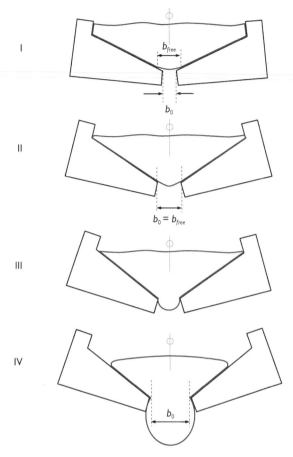

Figure F.1 Stages of geotextile container deformation during the opening of the split barge. Where b_0 = opening width of the split barge [m] and b_{free} = the opening width at which the container starts to slide to the opening [m].

L = length of the geotextile container [m];
φ = angle of shearing resistance between the floor of the split barge and the geotextile [deg];
θ = angle of the split barge floor to the horizontal [deg].

When there are many folds at the bottom of the geotextile, it is possible that the geotextile container immediately enters stages II or III, and bypasses stage I.

STAGE II

Stage II begins when the whole geotextile container begins to shift along the sides of the split barge towards its opening. Little deformation of the geotextile container occurs and the width of the suspended part remains fairly uniform. During this stage $b_0 \le b_{free}$ (see Figure F.1, stage II).

The tensile load acting in the geotextile at this stage depends on the fill material. Initially, if we consider the fill material to have zero internal friction (i.e. it is a liquid), the tensile load is governed by the friction between the geotextile and the sides of the split barge. The tensile load can be derived from an equilibrium consideration of the loads acting on the geotextile at the opening of the split barge [22]:

$$T = 0.5 \cdot \frac{\frac{W'}{L}}{\frac{\cos\theta}{\mu} + \sin\theta} \tag{F.2}$$

where:
μ = friction coefficient between the geotextile and the split barge [–].

For design purposes it is sufficient to increase the tensile load calculated using formula (F.2) by 10% to obtain the maximum tensile load that can occur during the opening of the split barge (stages I and II).

If the geotextile container is filled with a material that has internal friction (like sand) the tensile load in the geotextile is determined not only from the friction between the geotextile and the sides of the split barge but also from the shear stress that the fill material itself generates on the geotextile. In [22] the following formula is given:

$$T = F_h + 0.5 \cdot \frac{W'}{L} \cdot \frac{\mu \cdot \cos\theta - \sin\theta}{\cos\theta + \mu \cdot \sin\theta} \tag{F.3}$$

where:

$$F_h = 0.5 \cdot \gamma \cdot h_b^2 \cdot K_{\gamma a} - 2 \cdot c \cdot h_b \cdot \sqrt{K_{\gamma a}} \tag{F.4}$$

When the geotextile container is filled with sand the second term in formula (F.4) equals zero because sand has no cohesion (i.e. $c = 0$).

In formula (F.4):

$$K_{\gamma a} = \frac{1}{\tan^2(45° + \phi'/2)} = \tan^2(45° - \phi'/2) \tag{F.5}$$

where:
F_h = horizontal load in the axis of the geotextile container [kN/m];
γ = unit weight of the geotextile container [kN/m³];
h_b = height of the geotextile container (for closed split barge) [m];
$K_{\gamma a}$ = coefficient of active earth pressure [–];
ϕ' = internal friction angle of the sand fill [degrees].

Formula F.4 has been shown to generate conservative values for the horizontal load. The tensile load calculated is a safe design value but refinements can be made. Refer to [3] and [22].

STAGE III

The third stage begins when the opening width of the split barge is so large that the geotextile container can no longer be fully supported by the sides of the split barge, i.e. $b_0 \geq b_{free}$. The geotextile container begins to slide through the barge opening and significant deformation occurs. This stage ends when the barge opening width is large enough that the geotextile container in total starts to move downwards ($b_0 \leq b_{slide}$) with no further opening of the split barge ($db_0/dt = 0$). To determine the maximum tensile load acting on the geotextile during this stage an equilibrium model has been established [26]. This model has been fully explained in [3]. The tensile load on the suspended part of the geotextile can be calculated using formulae F.6 and F.7:

$$q_s = \frac{b_0 \cdot \gamma'}{2 \cdot \lambda \cdot f_g} \left[1 - e^{\left(-\frac{2 \cdot h_b}{2} \lambda \cdot f_g \right)} \right] \tag{F.6}$$

For the horizontal load on the most central part of the fill material:

$$F_h = \frac{b_0 \cdot h_b \cdot \gamma'}{2 \cdot f_g} - \frac{b_0^2 \cdot \gamma'}{4 \cdot \lambda \cdot f_g^2} \left[1 - e^{\left(-\frac{2 \cdot h_b}{b_0} \lambda \cdot fg \right)} \right] \tag{F.7}$$

where:

q_s = vertical effective stress in the lowermost part of the geotextile container [kPa];
b_0 = opening width of the split barge [m];
f_g = friction coefficient [–] described as: $f_g = \tan \phi$;
λ = in-situ earth pressure coefficient [–]: $K_{\gamma a} < \lambda < K_{\gamma p}$;
$K_{\gamma a}$ = coefficient of active earth pressure [–];
$K_{\gamma p}$ = coefficient of passive earth pressure [–];
h_b = thickness of the sand layer above the opening of the split barge [m];
γ' = buoyant density of the sand fill ($\gamma - \gamma_w$) [kN/m³];
γ_w = unit weight of water [kN/m³].

Figure F.2 shows a diagram of the forces acting during opening the split barge.

At the beginning of stage III, the barge opening will be relatively small. The in-situ earth pressure coefficient is then equal to the active earth pressure coefficient. As the split barge opens further, the horizontal stress will increase (the shape of the bin forces the sand to compress while being pushed downwards). At the end of stage III the horizontal stress, and thus the in-situ earth pressure coefficient will be larger, but also the opening of the split barge will be wider. It must be examined which of these conditions leads to the greatest loads on the geotextile.

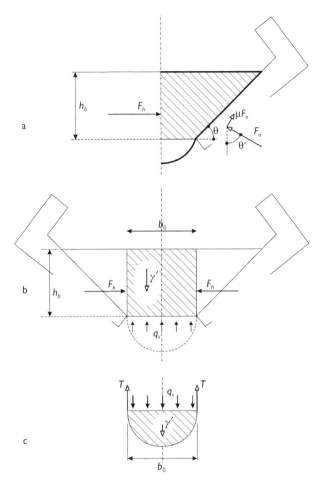

Figure F.2 Forces acting during the opening of the split barge.

To calculate the tensile load in the geotextile container the following formula can be used:

$$T = \frac{\pi}{16} \cdot b_0^2 \cdot \gamma' + 0.5 \cdot b_0 \cdot q_s \tag{F.8}$$

STAGE IV

The fourth stage can only occur if the circumference of the geotextile is relatively large and the further downward movement of the geotextile container is temporarily halted. After some deformation of the geotextile container a new equilibrium condition is reached. If the split barge is opened further, the geotextile container again begins to move down through the opening. This process can continue until the critical opening width is reached ($b_0 = b_{crit}$). An estimation of the critical opening width (b_{crit}) can be made by assuming that the geotextile container has a rectangular shape as it exits the barge. Thus, the opening width necessary for the container to pass through the barge opening is:

$$b_{crit} = \frac{S - \sqrt{S^2 - 16 \cdot A}}{2 \cdot A} \tag{F.9}$$

where:

S = circumference of the geotextile container (perpendicular to its axis) [m];

A = filled cross-sectional area of the geotextile container (perpendicular to its axis) [m²].

A more accurate method to calculate the critical opening width is given in [17].

If the geotextile container remains suspended for a brief time in the bin just before complete release from the split barge, the strength of the geotextile must be sufficient to carry a major part of the submerged weight. The tensile load in the geotextile is then nearly half of the submerged weight of the geotextile container per meter length. Due to the fact that the total amount of sand will not be situated at the bottom of the barge and thus not all sand will be in the lower part of the geotextile container, the formula is written as follows:

$$T = 0.45 \cdot \gamma' \cdot A \tag{F.10}$$

For example, if the container cross-sectional area is 10 m² and the buoyant unit weight of the fill is 9 kN/m³, then the load in the geotextile will be 41 kN/m. If the container remains suspended on one side of the bin for an additional brief period of time the load will be uneven and thus the tensile loads in the geotextile locally can be considerably greater.

Fall velocity of the geotextile container

When the geotextile container impacts the bottom, a significant load is exerted on the geotextile. This load must be quantified as best as possible because it is often the critical factor in determining the minimum required tensile strength of the geotextile.

An important parameter in determining the load upon impact on the bottom is the fall velocity of the geotextile container, which can be determined analytically using the following formula [1]:

$$v = v_\infty \cdot \frac{1 - e^{2 \cdot b'} + e^{b'} \cdot \sqrt{e^{2 \cdot b'} - 1}}{e^{2 \cdot b'} - e^{b'} \cdot \sqrt{e^{2 \cdot b'} - 1}} \tag{G.1}$$

where:

$$v_\infty = \sqrt{\frac{2 \cdot V}{A \cdot C_d} \cdot \frac{\rho - \rho_w}{\rho_w}} \tag{G.2}$$

$$b' = \frac{h}{v_\infty \cdot t'} \tag{G.3}$$

$$t' = \sqrt{\frac{2 \cdot V}{A \cdot C_d} \cdot \frac{\left(\rho + \frac{1}{2} \cdot \rho_w\right)^2}{\rho_w \cdot (\rho - \rho_w) \cdot g}} \tag{G.4}$$

where:
v = fall velocity [m/s];
v_∞ = terminal fall velocity [m/s];
b' = dimensionless water depth [–];
h = water depth [m];
V = volume of the geotextile container [m³];
A = maximum plan cross sectional area (in the horizontal plane) [m²];
C_d = drag coefficient (approximately = 1) [–];
ρ = density of the geotextile container fill [kg/m³];
ρ_w = density of water [kg/m³];
t' = characteristic time (=time necessary to reach 76% of the terminal velocity) [s];
g = acceleration due to gravity [m/s²].

The formulae given above show that the ultimate fall velocity of the geotextile container is a function of the density of the fill material, the drag coefficient and the shape of the geotextile container. In [22] the fall velocity is determined more accurately using a numerical model.

The formulae presented above do not lend themselves to a quick calculation by hand although they can be readily incorporated into a spreadsheet program. To demonstrate the influence of the different variables several design charts have been generated showing the relationships between different container shapes (relationship between volume and plan cross sectional area) and different fill materials. First, the relationship with terminal fall velocity is shown (Figure G.1) and then the relationship with characteristic time (Figure G.2). With these two values along with the actual

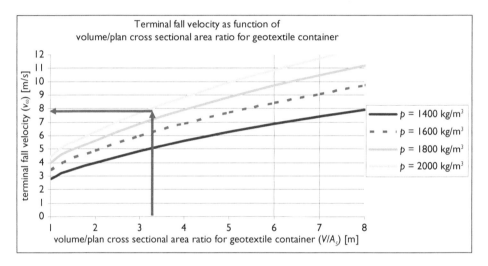

Figure G.1 Design chart for terminal fall velocity.

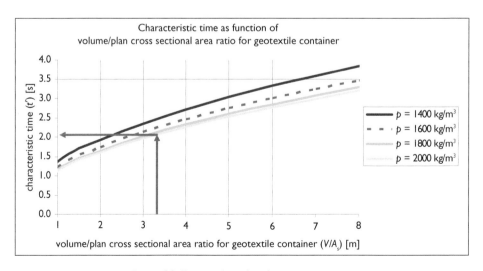

Figure G.2 Design chart for characteristic time.

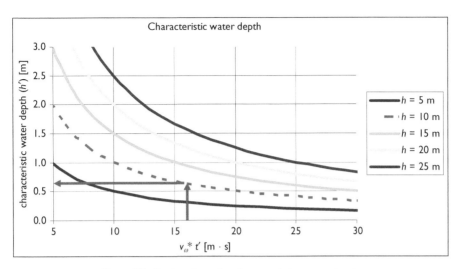

Figure G.3 Design chart for characteristic water depth.

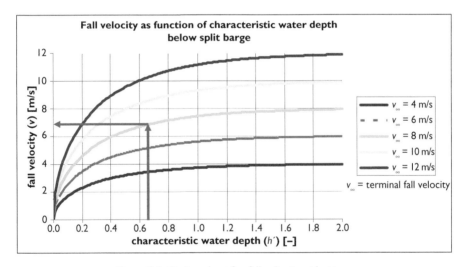

Figure G.4 Design chart for fall velocity at bottom.

water depth the dimensionless water depth ratio can then be determined (Figure G.3). Following this, the characteristic water depth and the terminal fall velocity can be determined, and the fall velocity of the geotextile container just before impact on the bottom (Figure G.4). This fall velocity ultimately governs the tensile load generated in the geotextile (Figure G.5), on the basis that all of the fall energy is absorbed by the geotextile. This last assumption, however, is a very conservative approach because part of the energy will be absorbed by the sand-fill, see Chapter 6.

In a field study (Kandia dam, Arnhem, The Netherlands.) the fall velocity for several geotextile containers was measured. The theoretical fall velocity for the containers

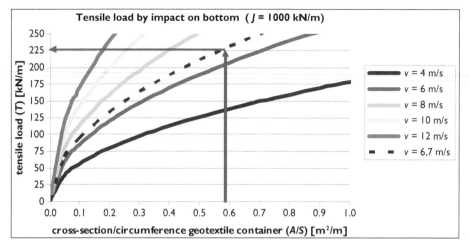

Figure G.5 Design chart for required tensile strength in geotextile.

was determined using formulae (G.1) to (G.4). A comparison between the measured and the calculated fall velocities showed that formulae (G.1) to (G.4) (with $C_d = 1.0$) are sufficiently accurate to be used in designing structures with geotextile containers.

CALCULATION EXAMPLE

For illustration, the calculation example from Chapter 6 is expanded here. First, the fall velocity and the corresponding tensile load generated in the geotextile are determined using the formulae. Then the same exercise is repeated using the design graphs.

$$v_\infty = \sqrt{\frac{2 \cdot V}{A_s \cdot C_d} \frac{\rho - \rho_w}{\rho_w} \cdot g} = \sqrt{\frac{2 \cdot 320}{100 \cdot 1.0} \cdot \frac{1990 - 1000}{1000} \cdot 9.81} = 7.9 \text{ m/s}$$

$$t' = \sqrt{\frac{2 \cdot V}{A \cdot C_d} \cdot \frac{(\rho + \frac{1}{2} \cdot \rho_w)^2}{\rho_w \cdot (\rho - \rho_w) \cdot g}} = \sqrt{\frac{2 \cdot 320 \cdot (1990 + 0.5 \cdot 1000)^2}{100 \cdot 1 \cdot 1000 \cdot (1990 - 1000) \cdot 9.81}} = 2.02 \text{ s}$$

$$h' = \frac{h}{v_\infty \cdot t'} = \frac{10}{7.9 \cdot 2.02} = 0.63$$

$$v = v_\infty \frac{1 - e^{2 \cdot h'} + e^{h'}\sqrt{e^{2 \cdot h'} - 1}}{e^{2 \cdot h'} - e^{h'}\sqrt{e^{2 \cdot h'} - 1}} = 7.9 \cdot \frac{1 - e^{1.26} + e^{0.63}\sqrt{e^{1.26} - 1}}{e^{1.26} - e^{0.63}\sqrt{e^{1.26} - 1}}$$

$$= 7.9 \cdot \frac{1 - 3.525 + 1.878 \cdot \sqrt{3.525 - 1}}{3.525 - 1.878 \cdot \sqrt{3.525 - 1}} = 6.7 \text{ m/s}$$

$$T = v \cdot \sqrt{\frac{A \cdot \rho \cdot J'}{S}} = 6.7 \cdot \sqrt{\frac{10 \cdot 1990 \cdot 10^6}{18}} = 233.000 \text{ Nm} = 223 \text{ kN/m}$$

The red lines on the graphs (Figures G.1 to G.5) are shown above demonstrating the same results.

Placement accuracy for geotextile containers

The placement accuracy is based on an average lateral displacement along with a corresponding lateral standard deviation. The magnitude of the average placement variability can be largely compensated during installation by taking appropriate precautionary measures, such as upstream dropping. The lateral standard deviation is random.

Small-scale tests at Deltares studying the placement accuracy for a drop depth of 15 m on a scale 1:20 showed rather large lateral displacement, see Table H.1, especially when the containers were dropped in water with a current. Theoretical studies confirm that such displacements are possible. In [20, in Dutch] the results of both the model study and the theoretical modelling are described.

The large horizontal displacements are probably the result of a smooth bed, which can form a water cushion between the bed and the (nearly fully descended) geotextile container. The geotextile container acquires a little lift from this cushion and the horizontal current velocity increases along with the horizontal displacement. When the bed is rough there is much less chance of a water cushion forming.

Based on the substantial horizontal displacements found in the Deltares model study, it was decided to perform field studies to analyse the placement accuracy data during the construction of the Kandia dam. A sonar scan was carried out both before and after the dropping of each container so that the position of the geotextile container on the bottom could be accurately determined and compared with the position of the split barge at the moment of drop. In [4] a design formula is presented, based

Table H.1 Average lateral displacement [m] (difference between horizontal drop position and position where the container comes to rest) and lateral standard deviation [m] for the dropping of geotextile containers at various current velocities and significant wave heights at a fall height of around 16 m (recalculated from small-scale tests). The large standard deviations in these tests lead to an additional study. See also text.

	Average lateral displacement		Lateral standard deviation	
	$H_s = 0\ m$	$H_s = 1.2\ m$	$H_s = 0\ m$	$H_s = 1.2\ m$
$u = 0$ m/s	0.8	4.6	0.8	4.0
$u = 0.5$ m/s	10.6	16.6	6.4	7.0
$u = 1.0$ m/s	22.6	24.6	3.6	1.2

on the field data measurements at a water depth of 15–22 m for still water without current (the average horizontal displacement is then zero):

$$s_p = 0.4 \cdot h - 3.2 \; (h > 8 \text{ m}) \hspace{4cm} \text{(H.1)}$$

where:
    s_p = standard deviation of the placement accuracy [m];
    h = water depth [m].

This result shows a somewhat greater variability for the situation where full-scale geotextile containers are placed than in the model study by Deltares. These results demonstrate that accurate placing on a sandy bottom at larger water depths is limited.

In the literature, model research has been carried out (scale 1:60) into the placement accuracy of geotextile containers in water currents [34]. The water current direction was parallel to the container placement. The drop depth varied from 30 to 48 m, current velocities varied from 0.5 to 2.9 m/s (prototype values). The authors describe that geotextile containers with a length of up to 15 m (which is relatively small in relation to what is common) oscillate some-what in their descent through the water. For elements in the shape of sandbags, the fall was of a spiral in nature. It was assumed that all the variability in placement was caused by the water current and thus the relationship between the theoretical displacement of x m by the water current and the placement accuracy as a standard deviation is given by $0.25 \cdot x$ m.

The placement accuracy of geotextile containers can also be compared to that of rocks dropped in water. This has been studied by Van Gelderen and Vrijling [28]. They observed that for several rocks the placement accuracy can be compared with a Rayleigh distribution where around 40% of the rocks have a standard deviation that can be stated as:

$$s_p = c \cdot \sqrt{h \cdot D_{50}} \hspace{4cm} \text{(H.2)}$$

where:
    s_p = standard deviation of the placement accuracy [m];
    c = constant (approximately = 0.7) [–];
    h = water depth [m];
    D_{50} = nominal diameter of the rock [m].

Although this formula is derived on the basis of rubble experiments and not for geotextile containers, it is important to note that according to this formula the standard deviation of the placement accuracy increases with increasing D_{50}, thus a large rock has a relatively larger potential deviation. For geotextile containers the deviation can also be considerable. The D_{50} value of a geotextile container is difficult to determine but it is certainly several metres (3 m or so). According to formula (H.1) a fall depth of 20 m would result in a standard deviation of 4.6 m (this corresponds well with the value found for the deviation using formula (H.2) of 4.8 m).

Theoretically, it can be demonstrated that the placement variability increases as the flow resistance becomes more influential. Current observations suggest the flow resistance only exerts an influence when it is comparable to the gravity forces that drive the fall of the geotextile container. As long as the container accelerates downwards, the flow resistance forces are still so small that they exert very little influence. When the container arrives at the bed the flow forces increase and create horizontal deviations; $h' = 1$ corresponds to a water depth of around 6 m (under the split barge), see formula G.3 in Appendix G. For rubble the terminal fall velocity is reached at a much lower water depth (depending on the size of the rubble).

In conclusion, the different studies referenced in this appendix show that placement variability can be a realistic phenomenon when placing geotextile containers at greater depths (more than 15 m). It is therefore advisable to perform field dropping tests at such locations where greater water depths are encountered and determine the placement accuracy by means of sonar measurements before and after the drops.

As stated already, the accuracy of placement of geotextile containers is very dependent on the water depth, the flow velocity and the wave height. From a practical perspective, the impact of these conditions on placement accuracy, is demonstrated by the following:

- Water depth to 10 m; current velocity to 0.5 m/s; significant wave height max. 1.2 m; geotextile container circa 300 m³. Slope of 1:1.5 possible with stacking of four containers;
- Water depth 15 to 20 m; no current; no waves. Evenly filled container horizontal deviation up to 3 m. Unevenly filled container horizontal deviation up to 5 m. Slope of 1:2 possible;
- Water depth circa 15 m; flow velocity to 0.5 m/s; significant waveheight max. 1.2 m; bed flat. In this situation horizontal displacements of max. 25 m measured.

A well-planned drop strategy based on the experience of the contractor can contribute considerably to the end result.

In summary, it can be anticipated that the placement accuracy increases when:

- the lower the water depth;
- the lower the wave height and/or flow velocity;
- the larger the filled cross sectional area of the geotextile container and split barge;
- the greater the density of the geotextile container [1];
- the more evenly the geotextile container is filled (with homogeneous sand, in terms of grain-size distribution and water content, applied layer by layer);
- the faster the split barge is opened;
- the rougher the bed surface.

References

1. Adel H. Den, (1996), Alternatieve systemen; geotextile container, GeoDelft rapport CO-365930, Delft.
2. Berendsen E. (1999), Project Maasvlakte 2, Haalbaarheid van de toepassing van geozandelementen, Waterbouw Innovatie Steunpunt, Utrecht.
3. Bezuijen A. (1999), Geo-systems, GeoDelft rapport CO-383990/8, Delft.
4. Bezuijen A., H. Adel, M.B. de Groot, K.W. Pilarczyck (2000), Research on geotextile containers and its application in practice, Proc. 27th ICCE, Sydney.
5. Bezuijen A., M.B. de Groot, M. Klein Breteler, E. Berendsen (2004), Placement accuracy and stability of geotextile container, Proc. 3rd European Geosynthetics Conference EuroGeo 3 pp. 123–128, München.
6. Bezuijen A., O. Oung, M. Klein Breteler, E. Berendsen, K.W. Pilarczyk (2002), Model tests on geotextile container, placement accuracy and geotechnical aspects, Proc. 7th conference on Geosynthetics, Nice.
7. Bezuijen A., R.R. Schrijver, M. Klein Breteler, E. Berendsen, K.W. Pilarczyk (2002), Field tests on geotextile container, Proc. 7th conference on Geosynthetics, Nice.
8. British Standard BS 8006: 1995 (1999), Code of practice for strengthened/reinforced soils and other fills. British Standards Institution, UK.
9. Buyze J.G., A.R. en Schram (1990), Stabiliteit van [grondkrib = artificial sill en onderwatergolfbrekers opgebouwd uit zandwordsten. TU-Delft, Faculteit der Civiele Techniek, Afstudeerverslag, Delft.
10. CUR publicatie 190 (1997), Kansen in de civiele techniek, deel 1: Probabilistisch ontwerpen in theorie, CUR, Gouda.
11. CIRIAC683 (2006), The rock manual, The use of rock in hydraulic engineering. 2nd Edition, London.
12. CUR/NGO publicatie 174 (1995), Geotextielen in de waterbouw, CUR, Gouda.
13. Palmerton J.B. (1998), SOFFTWIN-simulation of fluid filled tubes for windows. Computer Program. Copyright 1998.
14. CUR publicatie 214 (2004), Geotextiele zandelementen, CUR, Gouda.
15. De Groot M.B., M. Klein Breteler, A. en Bezuijen (2003), Resultaten Geotextile container Onderzoek.
16. Delft Cluster Rapport DC1-321-11, juni, Delft.
17. Groot M.B. de, A. Bezuijen (1999/2000), Designing with geotextile container, inventory for future research, GeoDelft report CO-391960/8, Delft.
18. Groot M.B. de, A. Bezuijen, A.K.W. Pilarczyk (2000), Forces in geotextile container geotextile during dumping from barge, Proc. Eurogeo 2, Bologna.
19. Klein Breteler M. et al (1994), Stability of geotextile tube and geotextile container, WL | Delft Hydraulics report H2029, Delft.

20. Klein Breteler M., R.E. Uittenbogaard, W.D. Eysink, M.B. de Groot M.B. (2001), Dump van geotextile container in stroming en golven, Modelonderzoek en numerieke simulatie, Delft Cluster rapport 03.02.01-05, WL | Delft Hydraulics H3679.10/H3820, Delft.

21. Leshchinsky D., O. Leshchinsky, Hoe I. Ling, Paul A. Gilbert (1996), Geosynthetic tubes for confining pressurized slurry: some design aspects, Journal of Geotechnical Engineering, ASCE.

22. Pilarczyk K.W. (2000), Geosynthetics and geo-system in hydraulic and coastal engineering, A.A. Balkema, Rotterdam.

23. Pilarczyk K.W. (1990), Coastal protection, A.A. Balkema, Rotterdam.

24. Schiereck G.J. (2004), Introduction to bed, bank and shore protection, DUP Blue Print, Delft.

25. Sylvester R. (1990), Flexible Membrane units for breakwaters, in 'Handbook of Coastal Engineering', John B. Herbich, ed., Vol. 1, pp. 921–938

26. Tsunoda N. (1995), Description of the dumping process, Mitshubushi Kagaku Sanshi Corporation, *private communication between Tsunoda and Pilarczyk*.

27. Broos E.J., P.M.N. van Zijl, W.J. Vlasbom, J.G. de Gijt, J. de Boer (2005). The design history of a promasing construction technique. Proceeding of the international conference on port-maritime development and innovation (pp. 1–10). Rotterdam, Havenbedrijf Rotterdam.

28. Van Gelderen P., H. Vrijling, W. Tutuarima (2000), The falling process of rubble dumped by a barge, Proc ICCE 2000 pp. 3920–3933, Sydney.

29. Van Rhee C. (2002), On the sedimentation process in a trailing suction hopper dredger, Proefschrift TU, Delft.

30. Venis W.A., (1968), Closure of estuarine channels in tidal regions, Behaviour of dumping material when exposed to currents and wave action, De ingenieur, 50.

31. Waal J.P. de (1998), Alternatieve open facing of a bank, Grondmechanica Delft/WL | Delft Hydraulics, rapport H1930, Delft.

32. WL | Delft Hydraulics (1973), Breakwater of concrete filled hoses, WL | Delft Hydraulics, Modelonderzoek M1085.

33. Bezuijen A., K.W. Pilarczyk (2012), Geosynthetics in hydraulic and Coastal Engineering: Filters, revetments and sand filled structures. Proc. EuroGeo 5, Valencia, pp. 65–80.

34. Zhu L., J. Wang, N.-S. Cheng, Q. Ying, D. Zhang (2004), Settling distance and incipient motion of sandbags in open channel flows, Journal of Waterway, Port, Coastal and Ocean Engineering, ASCE, March/April.

35. Heibaum M. (2004), Geotechnical filters – The important link in scour protection, 2nd international conference on scour and erosion, Singapore.

36. Klein Breteler M., C. Stolker, M.B. de Groot (2003), Final studies Geocontainer research, Delft Cluster Report.

37. Steeg, van, P.M. Klein Breteler (2008), Large scale physical model tests on the stability of geocontainers, Deltares Report H4595.

38. Steeg, van, P.M., E.W. Vastenburg (2009), Large scale physical model tests on the stability of geotextile tubes, Deltares Report 1200162.

39. Bezuijen A., H.J. Köhler (1996), Filter and revetment design of water imposed embankments induced by wave draw-down loadings, proceedings Eurogeo, Balkema Rotterdam.

40. Bezuijen A., P. van Steeg (2009), Geotextiele tubes en containers getest in de Deltagoot van Deltares, Geokunst, pp. 58–60.

41. Steeg, van, P., E.W. Vastenburg, A. Bezuijen, E. Zengerink, J. de Gijt (2011), Large-scale physical model tests on sand-filled geotextile tubes and containers under wave attack, proceedings 6th International Conference on Coastal Structures, Japan.

42. Lawson C. (2010), Geotextile containment, Proceedings 9th International conference on Geosynthetics, Brazil.

43. Leshchinsky D., O. Leshchinsky (1996), Geosynthetic confined pressurized slurry (GeoCoPS): Supplemental notes for version 1.0, Tech. Rept. CPAR-GL-96-1, USACE Watterways Experiment Station, Vicksburg, MS.

44. Timoshenko S., S. Woinowsky-Krieger (1959), Theory of Plates and Shells, McGraw-Hill Book Company, New York.

45. Cho S.M., B.S. Jeon, S.I. Park, H.C. Yoon (2009), Geotextile Tube Application as the Cofferdam at the Foreshore with Large Tidal Range for Incheon Bridge Project, Geosynthetics in civil and environmental engineering, Part 8, pp. 591–596.

46. Yee T.W., J.C. Choi, E. Zengerink (2009), Revisiting the geotextile tubes at Incheon Bridge Project Korea, proceedings 9th ICE, London, pp. 1217–1220.

47. Aminti P.L., E. Mori, P. Fantini (2009), Submerged barrier for coastal protection application built with tubes in geosynthetics of big diameter in Tuscany, proceedings 9th ICE, London, pp. 1235–1240.

48. Nikuradse J. (1933), Stomungsgesetz in rauhrenrohren, vdi-forshungsheft 361. (Engish translation: Laws of flow in rogh pipes, 1950. Technical report, NACA Techical Memo 1292. National Commission on Aeronautics, Washington DC).

49. Fürböter A. (1961), Über dieFörderung von Sand-Wasser-Germischen in Rohrleitungen (in German), Mitteilungen des Fransius-Instituts 19, 45, pp. 163–166.

Subject index

Abrasion 15, 60
application standards 17, 108, 110
armour stone 3, 4, 8, 14, 20, 24, 51, 103

barge opening 72, 125, 130
Bezuijen effect 83
blinding 103
blocking 5, 103
breaker parameter 24, 45

CE marking 17, 107
CEN standard 102, 110
Clogging 5, 103
concentrated acids 15
concentrated alkali 15
container deformation 126
cost 3, 20, 41, 60, 65, 78
creep 7, 8, 14, 15, 60, 62, 77, 109

damage during installation 3, 10, 14, 109
design approach 1, 4, 37, 40
design chart 6, 113, 114, 124, 132
design methodology 3
detergents 15
deterministic method 7
drainage 17, 37, 57, 61, 91
drop depth 135
drop strategy 137
durability tests 111

EN ISO 11058 102, 109

fall velocity (grains) 55, 56
fault tree 5, 8, 9
filtration 10, 11, 17, 108
friction angle 29, 43, 46, 113, 127

GEOCOPS 58, 62, 119, 121
geometrically closed filter 10
geotechnical stability 5, 46, 60, 113
geotextile bags 2, 4, 8, 19

calculation example 23, 32, 36
construction aspects 31
current flows 27, 29
failure modes 22
fall velocity 23
friction angle 29, 32
geometric design 21
groynes 19, 28
impact stresses 23
installation geometries 25
internal sand movement 29
land-based 20
liquefaction 22, 30
material choice 23
nominal thickness 22
overtopping currents 29
placement 20, 23, 25, 32
rupture of the geotextile 22
safety 22, 24, 35
safety considerations 22
seam strength 22
shear failure 30
slope factor 28
stability in waves 24
subsoil 28
tensile strength 23
underwater dams 29
uppermost elements 26
Geotextile containers 2, 8, 13, 19, 71, 133, 135
bin length 75, 76
bin width 76
Botjes zandgat 71
bund 88
caterpillar mechanism 86
Cornelis Douwes canal 71
cross-sectional area 73, 76
cyclic load 91
degree of filling 73, 78, 86
design formulae 86, 89
drop depth 135

T - #0624 - 071024 - C168 - 246/174/9 [11] - CB - 9780415621489 - Gloss Lamination